First edition

HAND RAISED

The Barns of Montana

Chere Jiusto & Christine W. Brown

Photographs by Tom Ferris

MONTANA HISTORICAL SOCIETY PRESS
HELENA, MONTANA

This project has been partially funded through grants from Humanities Montana, an affiliate of the National Endowment for the Humanities; the Montana Cultural and Aesthetic Trust; and the Preserve America program of the National Park Service, Department of the Interior, administered by the State Historic Preservation Office of the Montana Historical Society, which receives federal financial assistance for identification and protection of historic properties. Under Title VI of the Civil Rights Act of 1964, Section 504 of the Rehabilitation Act of 1973, and the Age Discrimination Act of 1975, as amended, the U.S. Department of the Interior prohibits discrimination on the basis of race, color, national origin, disability, or age in its federally assisted programs. Any views, findings, conclusions, or recommendations expressed in this book do not necessarily represent those of the National Endowment for the Humanities or Humanities Montana.

Humanities MONTANA

MONTANA CULTURAL TRUST
Partial funding for this project was provided by Montana's Cultural Trust.

NATIONAL ENDOWMENT FOR THE Humanities

Front cover photograph: Fisher Farmstead Barn, Park County
Cover and book design by Diane Gleba Hall
Typeset in Adobe Caslon and Calibri
Printed in Korea by Codra

Distributed by the Globe Pequot Press, 246 Goose Lane, Guilford, Connecticut 06437 (800) 243-0495

11 12 13 14 15 16 17 9 8 7 6 5 4 3 2 1

ISBN 978-0-9759196-9-9 (cloth)
ISBN 978-0-9801292-0-5 (paper)

LIBRARY OF CONGRESS
CATALOGING-IN-PUBLICATION DATA
Jiusto, Chere.
Hand raised : the barns of Montana / Chere Jiusto and Christine W. Brown ; photographs by Tom Ferris.

p. cm.

Includes bibliographical references and index.
ISBN 978-0-9759196-9-9 (hardcover : alk. paper) —
ISBN 978-0-9801292-0-5 (pbk. : alk. paper) 1. Barns—Montana—History. 2. Barns—Montana—Pictorial works. I. Brown, Christine W., 1972– II. Ferris, Tom, 1959– III. Montana Historical Society. IV. Title.

NA8230.J58 2011
728'.92209786—dc22
2010018795

For Jim, for always being with me,
and for Annika, Matt, and Laura,
you are all so fine to me.

Chere

For Ma and Dad, for never doubting,
and for Pete, Emmett, and Rose,
for sharing so much.

Christine

For Taylor, for her support and company, and for the barn owners, for keeping the barns standing and for their hospitality and directions!

Tom

The Montana Preservation Alliance and the Montana Historical Society would like to thank all those whose financial support helped to fund this publication.

Graham Foundation for Applied Studies in the Fine Arts

Humanities Montana, an affiliate of the National Endowment for the Humanities

LIATIS Foundation

Montana Cultural and Aesthetic Trust

Montana Department of Livestock

Montana Farmers Union

Preserve America program of the National Park Service

Thomas J. Teakle Trust at the Montana Historical Society

Treacy Foundation

And many generous individuals

Contents

Acknowledgments

WITH GRATITUDE, we acknowledge the people whose barns are featured in this book. They made this book possible, opening their barns and sharing their stories freely. We thank our friend Tom Ferris, whose photographs capture the beauty and convey the feelings that our words cannot, bringing the history alive and reminding us of the precious legacy that stands out on the horizon in every county and every back road in Montana.

This project would not have been possible without the unselfish efforts of many dedicated volunteers, colleagues, friends, and family members. We thank our workmates Patty Dean and Jim Jenks at the Montana Preservation Alliance (MPA); Kate Hampton and Pete Brown at the State Historic Preservation Office (SHPO); and MPA president Jim McDonald and our dedicated board members for unwavering support. We also thank our loyal volunteers Jim Greene, Miki Wilde, Annika Robbins, and Bobbi Kokoruda for their cheerful help in the office. In particular, we owe a great debt of gratitude to Jim Greene for his countless hours of research assistance on this book.

To the staff at the Montana Historical Society Press, Molly Holz, Glenda Bradshaw, and Christy Goll; to editors Jim Robbins, Rick Newby, and most especially, Ursula Smith; and to graphic designers Diane Gleba Hall and DD Dowden, we offer our heartfelt thanks for shaping our book, encouraging the project, and helping us to cross the finish line.

Many thanks to all the historians and museum curators, Extension agents, and humanists whose work inspired our book and who took time to find and share information with us. We especially thank Cindy Kittredge in Cascade, Susie Lindsay in Clancy, Vinola Squires in Dillon, and Ramona Raffaell in Sidney for time spent giving tours of their region's historic barns. We are also so grateful for assistance from Yvonne Gritzner and Humanities Montana and the community museums in Billings, Bozeman, Cascade, Clancy, Cut Bank, Dillon, Glendive, Great Falls, Hardin, Lewistown, Red Lodge, and Sidney that sponsored the "Barn Again!" exhibit and/or opened their collections of materials on barns or conducted tours that aided our research. The repositories at the University of Montana's Mansfield Library, special collections at Montana State University, and the Montana Historical Society Research Center were indispensable in our search for historic images. All who studied and wrote about barns and agricultural history before us blazed a fine path for us to follow. We also thank those who provided financial support for this work. We hope you all know how very grateful we are for this generous assistance.

Chere Jiusto and Christine W. Brown

• • •

AT MPA we take seriously our job to save and protect the places where the Montana saga was written. MPA has advocated for a quarter century for the places and landscapes that make the state unique. We work tirelessly to preserve the heritage that will belong to our children. It is a labor of love made stronger by the many members who join with us in this work.

Our colleagues at the Montana Historical Society and the Montana History Foundation are no less committed to this important cause. Together we published this book to document and celebrate the barns of our state, and we will continue on as partners to ensure that this heritage is preserved. While our project to compile this book is ending, our work to promote and advocate for Montana's rural buildings and traditions goes on.

You too can help. In a new partnership the Montana History Foundation, the National Trust for Historic Preservation, and the Montana Preservation Alliance have joined forces to create the Preserve Montana Fund, dedicated to preserving our state's historical treasures. Housed within the Montana History Foundation, the Preserve Montana Fund will help Montanans protect their heritage and the places that matter to them and will provide support for the repair of historic buildings, for collecting vanishing history, and for ongoing research and planning to safeguard the legacy of our fair state. We invite you and all who love barns to join our effort to preserve Montana's colorful past and these storied buildings.

MHF MONTANA HISTORY FOUNDATION

MPA MONTANA PRESERVATION ALLIANCE

NATIONAL TRUST FOR HISTORIC PRESERVATION®

INTRODUCTION

Montana's Barns

A Vanishing History

IN 1994, I landed in Washington, D.C., just in time for the opening of the National Building Museum and the inaugural of the Smithsonian Institution's "Barn Again!" exhibit. Inside the Great Hall of the museum stood a full-sized historic red barn. The barn had been dismantled and trucked to Washington and rebuilt on-site with an old-time barn raising, country fiddle music, and lots of elbow grease from two hundred volunteers.

No longer moored to its barnyard and pens and corrals, that barn looked like a ship out of water. There were no warm smells of animals and grain, no muffled stamping in the stalls, not even a scruffy old barn cat or two. But despite standing out of context, the barn was a comforting sight and a testament to the knowing hands that had first cut and fitted its timbers. Its walls bore the wear marks of decades of use—a 150-year-old grain barn that conjured up calving on cold winter nights, milking with the rising sun, and generations of people who had built their livelihood under the wood-shingled roof.

There is something about the ruggedness and weathered visage of an old barn and its unassuming elegance that stems from the wisdom and economy of purposeful design. Iron hasps forged by strong hands, native rock generously cut, hefty wooden planks sliced overly thick—these are the ingredients needed to render a barn sturdy enough to last and worthy of a long heritage.

Beyond their deep aesthetic beauty, barns also link us to the rural foundations of our culture—and to a time when so many more Americans were farm people. The notion that a citizen farmer might own property, have a voice in government, and exercise some control over his destiny still stands as a parable in the creation of this nation. Barns resonate with the power of this democratic allegory, and on some deep level, they inspire us to remember the importance of people coming together to build something big, something beautiful, and something that will preserve traditions for future generations.

• • •

THE "BARN AGAIN!" EXHIBIT was wildly popular, and the traveling version toured the country continuously for eight years, taking this celebration of rural life to the heartland of America. The images in the exhibit struck a deep chord: neighbors shouldering timbers and tall walls up into place . . . a shadowed figure walking the ridgepole on the skeleton of a new barn . . . stalwart red walls against rolling golden fields . . . soft, dappled light filtering into a silent haymow.

In a time when so much in our complex world is beyond our control, now more than ever the barn on the hill is a hopeful symbol. Our historic barns remind us that people in Montana have long been self-sufficient and resilient. Barns are an emblem of the importance of rural communities and family enterprise. They remind us that with spirit and ingenuity we can live in balance with the land, that we can support our neighbors who work hard and long to bring food to our tables, and that we can celebrate the bounty of their efforts. Those much-loved barns are an essential part of our cultural landscape; they are our heritage, and it is up to us to preserve them in order to honor the integrity of living in connection with the land.

• • •

RAISING A BARN was once a community event made possible by many hands. A hundred years later, preserving a barn can be almost as challenging. Historic barns are one of the most threatened historic resources in Montana. "Vanishing cultural icons" according to the "Barn Again!" curators, they are at risk for many reasons. In Montana, many barns were built during the boom years between 1880 and 1920; they are now 90 to 130 years old, an age when these wooden structures begin to wear out. Because of their sheer size, barns are expensive to maintain—to paint, roof, and repair. And they are outmoded—combines, pickups, and ATVs have taken the place of horses, plows, and wagons. Many sit in remote locations where rural life is fading and it is a challenge to keep businesses and families, let alone barns, in place.

Today, historic barns are disappearing from the horizon at an alarming rate. Just recently, the roof caved in on the MacDonald/Rossmiller Round Barn, one of only five round barns in Montana. Barn landmarks—the Doncaster Round Barn at Twin Bridges, the W. C. Child Ranch Barn near Helena, and the J. C. Adams Stone Barn outside Sun River—are critical to our heritage. But will they always be there?

This book was launched to help answer that question. Like grain elevators and beaverslide haystackers, barns are iconic in our Big Sky landscape. Yet our state has never undertaken a systematic survey of these buildings. With over six thousand historic barns noted in the 2000 census, we know there are heritage barns along every Big Sky byway and back road. A systematic project is desperately needed to document these barns and to preserve them *now*.

This book is not meant as a definitive history, and by no means is it comprehensive. Out of 6,000 barns, we have featured fewer than 140. And so, our book is instead an attempt to revel in the richness and the diversity, in the beauty and the dignity, of our heritage barns. More than anything, this book is a call to action. It is an effort to turn attention to these buildings and to raise interest in preserving them before they are gone.

CHAPTER 3
Western Valleys

CHAPTER 4
Great Divide

CHAPTER 5
Missouri River Country

CHAPTER 6
Yellowstone Basin

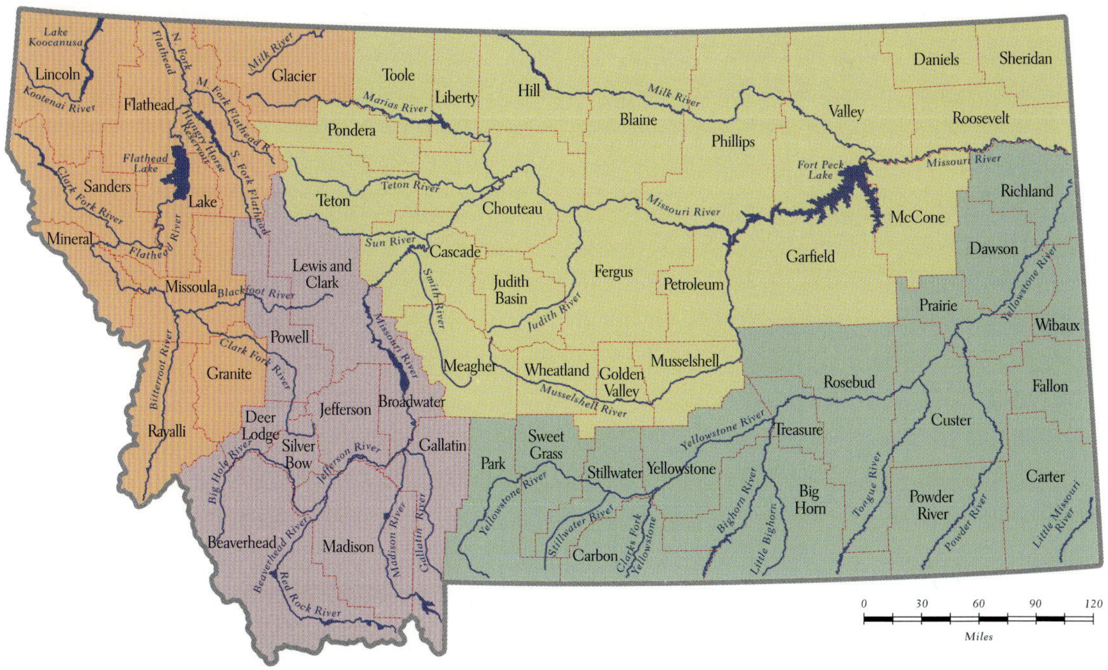

CHAPTER 1

Barn Building Traditions

IN 2003, the Beaverhead County Museum organized tours of some of the finest barns in that county, which was once the heart of Montana's young livestock industry. The tour explored some of the Beaverhead Valley's oldest barns, navigating the miles of gravel roads that connected far-flung cattle and sheep growers with each other and with the town and railhead of Dillon. Two of the barns on the tour were on the Blacktail Creek divide, summer range on the historic E. O. Selway Ranch, located twenty-five miles southeast of Dillon.

Now owned by the Matador Cattle Company, the Selway Ranch is part of its 300,000-acre Beaverhead Ranch, and the Selway Barns have been painstakingly restored by ranch manager Ray Marxer. The paired red barns that once held sheep and horses were built by John Kent, one of the valley's most talented barn builders. Among the guests on the tour were members of the first and the last families to have lived and ranched on that land—Jim Orr, a descendant of the Poindexter and Orr livestock dynasty, and Gayle and Katy Chaffin, who had ranched there from 1956 until 1987.

The Chaffins were accompanied that day by their daughter-in-law, Carleen, who related that electricity had not come to this area until the 1970s. Katy had raised the Chaffin children way out there frontier-style, thirty miles from town with no electricity and no phone. As the tour group scrambled up loading ramps, clicked photos, and took in all the handiwork that had gone into this four-story barn with its perfectly balanced wool lift and extra-sturdy draft horse stalls, the Chaffins stood quietly by. Carleen mentioned that this might well be the last time her aging in-laws would be able to visit the old place. Someone asked, "What do you remember most about living out here and using this fabulous barn?" The group might have expected "shearing the sheep" or "taking the horses through the mountains." The answer came after a long minute, in the plainspoken style of a Montana rancher: "I mainly remember that it was a lot of work."

Like the barn itself, Gayle Chaffin's answer captured the honesty and the reality of ranching life.

• • •

AGRICULTURAL barns originated in Europe as simple places for storing grain and sheltering animals. The word "barn"

(opposite) Selway Barn, Beaverhead County

descends from the old English *bereaern*: *bere* meaning barley and *aern* meaning house or store. It sometimes happens that a simple advance in technology changes much about the way people live. As a building for *aern*ing the *bere*, the *bereaern* became an essential part of year-round, settled agriculture; today, of course, the granary (from the Latin *granarium*) is in its purest sense the *bereaern*. And the buildings we know as barns now shelter livestock.

Sheltering livestock was critical in a northern climate, and over time a variety of buildings evolved to suit this need. Utilitarian from the ground up, barns were built according to tradition and purpose.[1] If there were windows, they were there to illuminate interior activities. If the barn was painted, that preserved the wood against weather and sun. If a cupola crowned the ridgeline, it served to vent the interior and keep green hay from molding, heating up, combusting, and burning the structure to the ground. Form followed function down through the ages, and with time, barn building evolved into a high craft, requiring individual training and know-how as well as the help of friends and neighbors when the time came to raise the roof beams.

A barn was often the biggest up-front investment that new settlers made. Many pioneer memoirs recount the tar-paper shack, the dugout shelter, or the one- or two-room log cabin that housed a family of eight or ten. Nearby was the barn, and here greater investment and faith were required. In many ways, the barn was more important than the house, for it sheltered the farm's wealth—cattle, sheep, and horses, along with crops and food.

From the Hands of Our Ancestors

In the American West, the intermingling of people from many nations created a rich tapestry of cultural life and led to a variety of building traditions. Ethnic groups that immigrated onto the High Plains brought with them styles of building and ideas on how to order the farmstead. It is possible to trace many architectural similarities from Old World to New. In Montana, the influences that shaped barn building traditions came from many directions—Anglo-American from the eastern seaboard, French Canadian from the North, ethnic European from the Old World. Across Montana, English-style barns, derived from colonial times in America, rubbed shoulders with double-crib "dogtrot" barns that traced a southern lineage, or with log *piece-sur-piece* barns reflecting French-Canadian heritage. Finnish and Norwegian log barns, masterfully hewn and tightly notched, belie the Nordic origins of many Montana homesteaders, while barns in the communities of Amsterdam and Manhattan tie back to Dutch ancestors. And in Lewistown and Columbus, Croatian masons carefully cut and stacked stone walls on barns that could, and did, stand for a century.

Barn design, however, was a dynamic process, and in America, barn building evolved over three hundred years as the climate, availability of building materials, changing animal husbandry practices, and applied knowledge of the builders all influenced the traditional design of barns. Architectural historians and cultural geographers often seek out the answer to the nature-versus-nurture question of whether rural architecture was more influenced by cultural patterning or by the regional environment. For Montana, one of the best studies on this topic was done by Terry Jordan, Jon Kilpinen, and Charles Gritzner, whose book, *The Mountain West*, offers insight into folk log building and the ways that many ethnic building methods were translated regionally. The authors note that in the West, while pure examples of ethnic barns can be found, there are more examples of those patterns having evolved as they were adapted to new regions.[2]

Other regional differences distinguish the barns of the West from those of the Anglo-European tradition. In some parts of Europe, stone, rammed earth, and brick barns are more common than they are here. There, centuries of harvesting lumber and clearing land left wood and timber in limited supply. In the heavily forested New World, as in western Montana, however, timber supplies were seemingly endless, and, compared with the time needed to quarry, shape, and lay stone, building with log was a speedy process. Thus, log and wood construction became prevalent here during the first decades of settlement.

Early Montana Barns

The **English barn** is the grandfather of the American barn. It was the English medieval threshing barn that became the prototype from which American barns descended. A symmetrical form, with a gable roof and doorways centered on the side under the eaves, the English threshing barn was three bays long with a storage crib on either side of a central alleyway that formed a threshing floor for wheat in the fall. With the doors open and a brisk breeze blowing through, grain could be tossed and the power of the wind harnessed to separate the wheat from the chaff. As John Michael Vlach noted in his comprehensive book *Barns*, English farms tended to have separate buildings for various animals and for storage. However, when the English barn came across the Atlantic, colonists, faced with harsher weather and limited time and resources, built the traditional English threshing barn but then filled it with cows, sheep, horses, feed, carts, and a variety of other things.[3]

The American version of this barn became common in the colonies and eastern Canada, then spread across the continent as the farming population moved westward. For the next hundred years, variations of the English barn remained common. Partitioned for keeping livestock and equipment on the ground floor, and for crop and hay storage on top, the gabled English barn was fairly simple to construct, modest in size, and practical.

Soon enough, colonial barn builders began to rearrange the traditional order of things. In eastern Tennessee, they relocated the doorway to the gable end in order to increase capacity. With the doorway on the end rather than on the side, the building could be lengthened along a central aisle, and far greater storage could be achieved when a full-sized loft was added. Long shed additions could also be added along one or both sides to enlarge the building or make space for milking rooms and other developments. The design took off across Appalachia and the South, and as the decades passed, **center-aisle**, or **transverse-crib**, barns grew larger. With the addition of higher, voluminous lofts, they accommodated ever more livestock and feed, and the format became quite popular as settlers moved into the Midwest and onto the American prairies.[4]

Log crib barns were commonplace during Montana's early history and fit well into the forested regions of the territory. Built of round or hand-hewn logs that were notched and joined at the corners, cribbed barn walls were stacked like Lincoln Logs, with each square forming a crib. These could be built singly to form a truly basic one-room barn, or, by simply extending a roof to connect two freestanding cribs, a double-crib, or dogtrot, barn could be created, providing a covered space between the log walls that could be used alternatively for threshing, calving, wagon storage, or a sleeping spot for Shep.

Many crib barns trace their lineage to the American South, to French Canada, and overseas to Scandinavia. While stacked log walls, a square crib layout, and a gable roof all represent this tradition, variations abound. The crib became a building block that could be combined into double- and four-crib designs. Cribs could also be strung together to form a long, low building for barns or residences.

The long and low multicrib log barn on the Sims-Garfield Ranch in Golden Valley County was built by stringing several single-crib barns together, a common pattern on the High Plains.

Montana Log Barns

At least two regionally distinctive log-barn types emerged as farmers and builders moved into Montana Territory and adapted to the local environment: the mountain horse barn and the multicrib log barn. One fits the forested western mountains of the state; the other is a low-profile, ground-hugging response to the exposed eastern prairies.

In *The Mountain West*, Terry Jordan, Jon Kilpinen, and Charles Gritzner describe the **mountain horse barn** as a type of vernacular barn building that evolved in the West. A single-crib barn built of logs notched at the corners and spanned by one continuous roof, the mountain horse barn could vary in size, but it was distinguished by its long log walls that created a large single crib on the interior; a central entrance but not always a through aisle; and a high hayloft. Doorways were located in the gable ends rather than on the side. Unlike many barns to the east, where hogs and dairy cattle were often the inhabitants, the mountain horse barn held horses and far more hay—a direct response to a harsher climate and the need to protect and feed horses through long, cold winters, while range cattle did not generally require such shelter.[5]

Western builders were designing such adaptations in response to regional conditions. The mountain horse barn, like the beaverslide haystacker, appears to have originated in southwestern Montana's Beaverhead country, a "hearth of invention" on the frontier of western high-valley cattle ranching.[6]

The **multicrib log barn**, in contrast, is a long, low building formed by stringing together several log cribs and topping them with a low-profile sod or stone roof. The multiple-crib design made the most of more limited wood supplies and could be added on to over time. There was no loft; grains and hay were either kept in a crib or in separate granaries. Both barns and cabins were created in this manner, and the long, sprawling log ranch buildings that resulted in central and eastern Montana, Wyoming, and elsewhere in the West are a folk building style deserving of more study.

Ethnic Barn Types

On the frontiers of settlement, where trees were abundant but milled lumber scarce, log buildings were the order of the day. They stood up to harsh weather and hard use. In the West, and in territorial Montana, there were many log builders who applied their skills to transform mining camps and early rural settlements. They left behind an array of log architecture that reflects the energy and diversity of those early communities.

French Canadians who followed the fur trade west from Ontario, Quebec, and the prairies of Saskatchewan brought log-building techniques with them. Tight log walls strung between vertical posts was a method of construction known as ***piece-sur-piece***. French-Canadian carpenters employed this technique to erect forts and buildings along the earliest western frontiers, and in Montana we see *piece-sur-piece* craftsmanship in rare early barns.

Nordic settlers like Swedish carpenter Carl Swanstrum in Beaverhead County brought log-working traditions from the deep forests of northern Europe (see Swanstrum Barn, page 96). Using hand tools to cut, shape, and fit the logs, these carpenters executed exacting corner notching and hewed log walls to fit tightly and to last.

One key to deciphering the origins of a log barn is the notching at the corners of the walls. The style of square, saddle, or dovetail notching can speak volumes about the craftsmen who cut and fitted the logs. There are many kinds of log joinery, and in places like Bannack, Virginia City, and Nevada City, where large collections of log buildings have been preserved, one can find a wide variety of notching and wall-construction techniques attributed to the varying traditions of the craftsmen who built them. Throughout Montana and the West, square-notched, V-notched, and saddle-notched barns show various ideas for joining logs together and holding them in place.

On Montana's arid eastern plains, a scarcity of timber and funds led settlers to literally scrape what they could from the prairies and to build with ingenuity using the sod they broke for their fields, dirt that they rammed into building bricks, and sometimes just holes dug into hillsides that they modified into buildings. "Soddies" (or sod homes), dugouts, and tar-paper shacks were symbolic of homestead settlement across the Great Plains. Rammed-earth "basta brick" buildings, often associated with German Russians, and sod-walled structures were built by homesteaders who formed a wave of farming settlement across the Great Plains and the Dakota prairies and edging into Montana.[7] These were temporary buildings, not intended to withstand a century or more of exposure to the elements. Few, if any, sod buildings remain in Montana today.

Many Europeans who immigrated to Montana brought masonry skills that derived from strong traditions of stoneworking. While there were many ethnic masons, Croatian and Italian stoneworkers stand out historically. Lewistown, Columbus, and Helena were areas of high Serbo-Croatian settlement, and the barns and stone buildings of these towns show off the skills of those masons. Their abilities to quarry, cut, and shape stone—from soft sandstones to much harder

In Helena, men of the Jezick family applied traditional Croatian stoneworking skills in the construction of schools and other civic buildings. They also built a stone home and a multifamily building, where, in 1900, fourteen Croatian stoneworkers lived. The barn, which still stands today, reflects their impressive skills.

Half dovetail, Boyer Barn, Missoula County

Piece-sur-piece, Lingshire Barn, Meagher County

Lap, Silver King Ranch Barn, Lewis and Clark County

Tooth, Heikkila-Mattila Barn, Cascade County

Full dovetail, Renlund Barn, Carbon County

Saddle, Ober Barn, Lake County

Notch Styles

Corner notching showcases the skill of barn builders through the ages.

Steeple or V, Grant-Kohrs Ranch Barn, Powell County

Square, Lingshire Barn, Meagher County

porphyry and granite—are impressive. In the Helena vicinity, Italian stone masonry stands out on the W. C. Child Ranch Barn at the White Face Farm (see page 128).[8]

From limestone to granite to travertine, the stone quarries in these locales all yielded an ample supply of stone for foundations and walls of many outstanding buildings. Showcasing masonry talents, the stonework on many barns also reflects the area's geology, from cut granite to river cobbles. And, of course, whether it was rubble stone, cobblestone, or rusticated, shaped, or dressed cut stone, all invested the barns of Montana with a strength and a beauty that surpass time.

American Barns

The traditional barns of America evolved over a period of three hundred years, and in doing so, they increasingly lost much of their ethnic character and regionality in favor of functional adaptations and design concepts. "In America there are two phases of barn design," wrote architectural historian Thomas Hubka in his essay "Americanization of the Barn." "If the story of the development of the American barn could be told in two parts, the first part would describe the transmigration of the European barn traditions and their spread and diffusion in America. The second part of the story would describe the Americanization of those European traditions into American national and regional types. Whether the story is described as the melting pot of various Old World ideas or attributed to the modernizing processes of American technological and economic development, the results are the same—the European barn became Americanized."[9]

"Americanizing" the barn took on a measure of velocity with the Industrial Revolution of the mid-nineteenth century and the advent of mass-produced materials and publications on design and efficiency. During the 1830s and into the 1850s, books by A. J. Downing and A. J. Davis expressed ideals for rural design and captured the attention of American readers. In the decades that followed, an organized approach to design was taken up by publishers of pattern books, purveyors of building kits, and agricultural programs in universities.

Things like straw thatch, slate, and tile roofs and mortise-and-tenon framing became increasingly obsolete as the railroads greatly increased the volume and array of available building supplies. New technologies were based on manufactured materials. Large quantities of milled lumber, for example, gave rise to the **plank-frame** or **balloon-frame** structures that Downing promoted, which featured a lighter wooden skeleton over which traditional wooden siding was applied. These open forms made it possible to build walls higher, expand lofts, and shape rooflines and still use less material overall than used in heavy-log construction. By the turn of the twentieth century, designer William Radford reported that builders had improved upon early experiments where barns not built strong enough were wrecked by heavy winds and that balloon-frame roofs were becoming "quite popular in barn construction."[10]

The rails also brought other materials, including glass, steel, concrete block, and bricks, which gradually infiltrated the building trades and changed the look of barns and other buildings. Helping to promote the revolution in barn design was the work of the nation's agricultural colleges. As agriculture moved into the twentieth century, barn building increasingly took on scientific ideals. Folk building methods gave way to standardized plans that incorporated new technologies and experimental research. Milled lumber, manufactured windows, concrete flooring, systematic written plans—all changed the way barns were designed and the way they were used.

Farmers' bulletins, widely available, carried plans for dairy, horse, sheep, and beef cattle barns, along with other buildings suited to rural life, including "modern" farmhouses and even doghouses. *Radford's Combined House and Barn Planbook*,

Under One Roof

The shape of a barn roof can tell a lot about its history.

Front gable, Lingshire Barn, Meagher County

Side gable, Cobb Barn, Park County

Conical, Fitzgerald Round Barn, Teton County

Hip, OW Ranch Barn, Big Horn County

Gambrel, Blackwood Barn, Gallatin County

Monitor, Bos Farm Barn, Gallatin County

Round roof, Wampler Barn, Park County

Saltbox, Waterman Barn, Gallatin County

Gothic arch, Bonnell Barn, Park County

"being a complete collection of practical, economical and common sense plans" for the industrious farmer of 1908, expressed the wisdom of the day: "Man, the king of all, merely follows the law of all creation in fashioning for himself beautiful habitations, and he goes a step further even in building beautiful structures for the housing of the horses, the cattle, the sheep and the other animals that are dependent upon his care and his bounty."[11]

Roofing styles varied through time with changing technologies and materials. In nineteenth-century Montana, most barn roofs were gable, monitor, or rarely, saltbox. The **gable roof** is the simplest, created when two opposing planes of the roof meet at a central ridgeline. On a barn, the gable runs parallel or perpendicular to the aisles that run through the building. The **saltbox form** is a variation, deriving from the East Coast, where heavy snowfall and Yankee ingenuity together shaped the design. Like the saltbox houses of New England, the saltbox barn has a steep roofline that descends from a high gable. The gable is off center, and the steep plane of the rear roof is longer and reaches closer to the ground than does the front. On a barn, this offered the advantage of economically expanding the usable interior space by simply extending the plane of the roof.

The need for interior lighting mothered the invention of the **monitor roof**. Before the widespread availability of electricity, simply raising the center ridge of a gable roofline and adding windows along the side walls allowed more natural lighting into a building. Besides their use on barns, we see these roofs used on industrial buildings across Montana—mills, foundries, and warehouses. Through the decades, many monitor barns were built across Montana, and the form is still used today.

The **gambrel roof** was developed in Europe and built in America from colonial times by ethnic builders using post-and-beam techniques. With the addition of a braced joint midway along the plane of the roof, the interior volume could be expanded. This provided an expansive upper storage space and far more capacity for the hay needed to keep livestock through the winter. The form took on more widespread popularity during the early 1900s when lighter milled lumber made it possible to move the roof's structural elements to the outer edges of the loft and leave the central space unobstructed.[12] The lyrical name "gambrel" comes from the medieval Latin word *gamba*, or the Norman *gamberel*, meaning leg of an animal, and indeed the shape of the roofline mimics the back leg of a horse.

One of the most distinctive roofline finishes is the **Gothic arch**, or the **pointed arch**, barn. The Gothic arch form begins at a peak in the roof and traces a curving profile that embraces a large loft where much hay and grain can be stockpiled through long, snowy winters. Building historians trace the Gothic arch to Michigan and the late nineteenth century. Rather than the straight rafter, the key to this roof form was in shaping the rafters along a curved pattern, by laminating several layers together to create a "bent" and then raising them into place like ribs along the interior of the loft. Plans for these barns came from both agricultural universities and building kits.

The **round** or **polygonal barn** with its conical roof also goes back to colonial times. George Washington himself was an innovator. In 1793, he designed a sixteen-sided brick threshing barn with a circular indoor alley where a team of horses would tread the grain. In 1826, the Shakers built a truly round stone barn in Massachusetts that was widely noted. Over the next century, round barns and those with eight to sixteen sides were constructed, but not on a large scale. In 1889, Franklin King designed a two-level round barn with a central silo for easy storage and feeding. And at the turn of the twentieth century, the round form became popular as agricultural programs, particularly those at the universities of Illinois and Wisconsin, promoted it.[13]

Round barns were designed for economy—in volume-to-surface area—and efficiency. In their day, they were lauded for the systematic way they allowed a farmer to move the cows into stanchions, feed all from a central hub, milk them, and easily clean the alleyway encircling the perimeter. Round-barn building know-how came to Red Lodge in 1941 when Emery L. McNamee built a round barn for the Kent family (see page 244). As brick walls for the Kent dairy were being laid, the barn caught the attention of a national farm and home radio hour, and its progressive features were discussed. Over the airwaves, McNamee explained the "manifold advantages" of barns in the round, not least of which was the fact that they were "nearly windproof with no corners to catch and hold the wind."

Montana never had many round barns; just a handful are known. The most famous one in Montana is the Doncaster Round Barn near Twin Bridges (see page 98). Built in 1886 for Noah Armstrong, the Doncaster Barn is the birthplace of Spokane, the Thoroughbred who in 1889 became Montana's only Kentucky Derby winner. Other round barns include the Fitzgerald Round Barn at Collins and the MacDonald/Rossmiller Round Barn east of Brady (see pages 156 and 160).

A Barn for Every Animal

While we know that barns descended from a long line of granaries and storage buildings, the stories of Montana's barns are largely about sheltering and feeding and caring for livestock. Dairy and horse barns are the largest and most impressive, but there were also buildings that provided for the care and handling of other animals. During the 1920s, Montana State College's agricultural extension program and Professor W. R. Plew created designs and plans for outbuildings to house all types of farm animals—from sheep and pigs to cattle and horses.

Montana had a few round barns, and a handful of them have survived. This interior view of the Fitzgerald Round Barn in Teton County shows the magnificent roof structure.

The James Manufacturing Company, a leader in technological improvement of dairy barns and equipment, described the advantages of plank frame construction in its 1917 plan book, pronouncing this kind of barn "so easy to build that a skilled carpenter is hardly required to do the work." Two of the framing phases for a plank frame barn are illustrated here.

Horse Barns

These barns featured stalls for indoor shelter in bad weather, extra-strong stalls and feed troughs for workhorses, heavy-duty stallion stalls, ample haylofts for feed storage, and tack rooms for keeping harnesses and gear. Extension sheds expanded the capacity to hold buggies, wagons, and blacksmithing and harness repair work all under one roof, while loafing sheds offered a place for livestock to get under cover and out of the weather.

Dairy Barns

When the cows came home to the barn, a feed trough invited them in. A clean milking parlor and a cool room for storing milk and cream, topped by grain bins and hay storage in the loft, were standard features. For more storage, larger dairies often included a silo or two.

Cattle Sheds

While most range cattle were left outside to their own devices, bull barns for prized breeding stock and indoor space for calving became desirable.

Sheep Barns

At one time there were many more sheep barns on the horizon, particularly in central Montana. Sheep ranchers devised ingenious systems for lambing, dipping, and shearing, including complex ramp systems that allowed sheep to run through, be counted, and be shorn. Central floors and stalls doubled as lambing enclosures, and often there were pens for shearing.

Chicken Coops

In the old days, chickens were raised mostly for family sustenance as well as for a little egg money. But today we see the whole range of coops in Montana, from the little coops housing chickens for a few eggs to the enormous chicken barns operated for commercial production, notably on Hutterite colony farms.

Pig Barns

A familiar sight in the Midwest and in some places in Montana, these structures keep a low profile, with their lines of nursery pens along the sides. There is also space for sows and for weaner pigs as they grow.

Specialized Barns

When harvest time came, there were barns for individual crops—potato barns, granaries, hay barns. These were essential for crop storage and specialized marketing. There were also barns for sales, for showing, and for fairs. Once the center of life for entire communities, these barns are still vital to the continued marketing and sale of livestock and the celebration of all things agricultural.

Form Follows Function

Barn design has been an evolving art for generations, and as technology advanced, elements of agricultural buildings—from silos to manure hoppers—were analyzed and improved. The architectural profession developed, and with it numerous talented designers came onto the scene. Some of Montana's best-known architects—Robert C. Reamer, Fred Willson, A. J. Gibson, Ole Bakke, Chandler Cohagen, H. E. Kirkemo—drew plans for barns, ag buildings, and farmhouses. Many of these buildings remain today, and some of the original architectural drawings are housed in state university archives.

Specialized designs also proliferated through companies like Sears, Roebuck and Company, Alladin, and Montgomery Ward, which all sold plans and kits for homes and farm buildings. Sears also sold complete building kits, with all materials delivered on-site.[14] Examples in Montana include the Jemison Ranch near Great Falls, which is an example of the picture-perfect Sears farmstead, with its sturdy Craftsman-style home and its gambrel-roofed, all-purpose barn (see Jemison Barn, page 172).

By the mid-twentieth century, even plan books and blueprints took a backseat to custom-designed barns drawn by teams of designers. One of the leading companies, the James Manufacturing Company (today, Jamesway Farm Equipment), founded in Wisconsin, designed both new barns and makeovers of earlier barns. The McHugh Dairy Barn in Helena, built circa 1920, was one of its buildings and was featured in the catalog.[15]

Today, the barns of many decades are still with us and are still serving useful purposes, while the world of agriculture continues to evolve. Technologies such as pole barns and steel-framed, tension-fabric structures are taking older barns' places on the horizon, but they will never match the craftsmanship, quality, and aesthetic beauty of the nineteenth- and early-twentieth-century barns. The most effective way to ensure that historic barns do not fall into the forgotten realms of history is to use them, to reinvent their purpose, and to reinvigorate their usefulness.

CHAPTER 2

Agricultural Settlement in Montana

THE SEEDS of modern-day farming and ranching in Montana were cast in the Bitterroot Valley in 1841, when missionaries founded St. Mary's Mission at the confluence of the Bitterroot and Burnt Fork of the Bitterroot rivers. That fall, Father Pierre-Jean De Smet and his Jesuit brothers introduced Montana's western Indian tribes to agricultural practices and crops that came from the Old World. The Salish, Pend d'Oreilles, and other Native people who called this valley home had highly sophisticated horticultural traditions; they harvested plant foods and medicines throughout the year. But the introduction of the farming of grains and produce and of domesticated livestock expanded the natural cornucopia of the region.

That first year, the missionaries built the mission and outbuildings from century-old cottonwood trees and broke the soil to plant crops. The setting was conducive, for the Bitterroot is Montana's most fertile valley, and nature is at its gentlest here, offering mild winters, abundant moisture, a long growing season, and dark, rich soils.

Sprouting up alongside the oats, wheat, and potatoes were a chapel, houses, and, among the mission's new outbuildings, log granaries to store the wheat and other grains. The Jesuits were talented builders, and their skills are evident here and at nearby missions such as St. Ignatius in the Flathead Valley and Sacred Heart at Cataldo, Idaho. Sketches by Father Nicholas Point and others at the missions provide a glimpse of Montana's early-day buildings for agricultural storage.

With forty-six head of cattle, along with hogs, St. Mary's stock became a foundation herd in the stock-raising history of Montana. To the north, in the Mission Valley, the Jesuits opened St. Ignatius Mission, where, by the late 1850s, they had introduced sheep and were running about a thousand head of cattle.[1] Thus, as the missions began to transform the lives of Native people and early settlers in these valleys, the first farmers began to transform the landscape of the Bitterroot and the surrounding territory.

Forts and Farming

Elsewhere, fur-trading posts and military forts strung along the major rivers and travel routes into and through the territory were home to Montana's first widespread agricultural endeavors. Small gardens and plots of staple vegetables such

(opposite) Inside the Staudaher Barn, Beaverhead County

The Jesuit priests of the Catholic Church were talented builders, their skills evident in the missions they built in the mid-1800s in Montana and Idaho, including "St. Mary's among the Flat-heads" (above), which appeared in Father Pierre-Jean De Smet's book describing his time in the Rocky Mountains in 1845.

as potatoes, corn, onions, cucumbers, peas, and beans helped to sustain these posts. During the 1830s and 1840s, a few dairy cattle were introduced at Fort William on the Yellowstone and at Fort McKenzie on the Missouri, near present-day Loma.[2]

In 1850, after St. Mary's Mission faltered, Major John Owen established a trading post there. Two dozen farmers and stockmen settled nearby, among them Thomas W. Harris, who was reportedly Montana's first full-time farmer, along with midwesterners and numerous Métis French Canadians. Owen's journals noted that Harris transported produce to market on a regular basis, and about 1860 Owen himself imported three hundred head of sheep into the valley. No more than a decade later, Montana boasted 851 farms spread over 130,000 acres of land.[3]

Whiskey for the Men, and Fresh Horses

Growing numbers of emigrants following the Oregon Trail sparked the nascent livestock industry in Montana. The emigrants' livestock were starving and exhausted by the time they reached the Rocky Mountains, having walked for months over trails increasingly stripped of grass by the tens of thousands of animals moving in a steady swath across the continent. The 450-mile stretch from Fort Laramie (Wyoming) to Fort Hall (Idaho) was particularly dry and grueling. It was about 1850 when Richard Grant, a Hudson's Bay Company trader, and his sons Johnny and James set up business at Fort Hall, trading at a rate of one fresh animal for two weary ones. Then, after trailing exhausted cows and horses from the Oregon Trail up into the Beaverhead and Deer Lodge valleys, the Grants wintered the animals over. In the spring, fattened on bunchgrass, blue gramma, buffalo grass, and wild fescues, the stock returned fresh for trade with the next wave of travelers. The Grants and other pioneering cattlemen in the Bitterroot, Deer Lodge, and Beaverhead valleys soon built the first big cattle herds in the region. By the late 1850s, beef cattle from these valleys were being marketed to the military and the fur traders, and by 1858, with six hundred head to his name, Johnny Grant established a ranch in the Deer Lodge Valley.[4]

Nearby, in the mountains surrounding the Deer Lodge Valley, a new era of settlement was about to dawn. Despite rumors of gold described by François Finlay and others in the early 1850s, the first recorded discovery was made in 1858 by Deer Lodge stockmen Granville and James Stuart. The brothers found nuggets amid the boulders of a small waterway they christened Gold Creek, and the strike marked the birth of modern-day Montana. As word spread, a growing crowd of hopeful miners detoured here from other western gold camps and from eastern states.[5]

The wealth from mining raised the appeal of western territories, and places that had seemed distant and empty began to look more alluring to prospectors, businessmen, and fortune seekers. News of strike after strike issued from the valleys of what was first Washington Territory (1853–1863), then Idaho Territory (1863), and finally Montana Territory (1864). The rush to Montana's goldfields was on.

Bustling mining camps at Bannack, Virginia City, Last Chance Gulch, and elsewhere created a brisk demand for foodstuffs. Mormon farmers freighted flour and produce up from Utah to the goldfields, and soon enough, farms and ranches sprang up in the high valleys of the Madison, Beaverhead, Bitterroot, and Gallatin rivers to supply the camps. Herds of livestock increased accordingly. As Merrill Burlingame and K. Ross Toole reported in their seminal work on Montana agricultural history, Montana's tax rolls for 1866 listed 1,896 cows, 464 mules and asses, 1,769 sheep, 249 hogs, and over 4,300 oxen. By the start of the 1870s, these numbers had doubled and tripled.[6]

Freighters harnessed the power of oxen to haul goods arriving by rail and steamboat into the territory and to move Montana's goods out to market. Mules and "bull teams" of oxen were used to pull long trains of wagons hitched end to end and loaded with freight. Wool was one of the main territorial exports, and through the 1870s it was freighted over Monida Pass to Corinne, Utah, or to Fort Benton for steamboat passage down the Missouri.[7]

Over the Plains, through the Mountains

Montana's population grew rapidly during this period, and cattle and sheep ranchers expanded their operations. Moving large herds of livestock across long distances and rugged terrain was a daunting challenge. Roads into the territory were primitive, the main transportation corridor being the wagon trail from Corinne, Utah, that ran through the Beaverhead Valley and on to Virginia City. It was not until 1860 that the first steamships chugged up the Missouri River to Fort Benton, and not until 1880 that the Utah & Northern rail line penetrated Montana.

"Last of the Bull Teams, 72 Bulls, 16 Wagons, 96,000 Pounds of wool, July 1892" declares the caption on this photograph taken in Big Timber. Sturdy oxen could pull heavier loads and negotiate rougher terrain than horses. And unlike horses, oxen rarely needed to be stabled.

The XIT Cattle Company became legendary for driving cattle to Montana from Texas, over a thousand miles away. Shown here, XIT cattle graze on open range on Big Dry Creek, a Missouri River tributary in eastern Montana, in 1904.

Until the railroad's arrival, the best option for moving livestock into the mining West remained buying cattle, horses, or sheep and then walking them to Montana. Many came from a "nearby" port like Puget Sound, Washington, or The Dalles in Oregon. In the mid-1860s, Philip Poindexter and William Orr drove four hundred steers from California to the mining camps of Virginia City and Bannack, while Nelson Story trailed six hundred Texas longhorns from Dallas to Bozeman.[8]

Trailing cattle to market was as difficult as bringing them into Montana. It generally involved a two-month journey south to Union Pacific rail line in Wyoming or Utah. The distances were vast, the trip arduous, and the prospect of tangling with foul weather, hostile Indians, or unexpected accidents made the journey extremely dangerous. In short, all the ingredients of a mythic American adventure are woven into the history of western livestock.

Many of the same challenges limited construction of frontier buildings. Thus, for the most part, those buildings were made of regional materials—log and rubble stone—before materials such as sawn lumber and cut stone were feasible or available.

Days of the Open Range

As lands that had belonged to tribes came into the federal domain as a result of treaties, livestock growers began to fill the ranges. Cattle and sheep were free to roam the rolling plains and open valleys as the buffalo had done for millennia. In the 1870s, ranchers spread into the unfenced Sun River Valley, and several thousand head of cattle were driven into Montana from Washington and Oregon.[9] Early outfits were soon joined by others looking to cash in on "easy money." Investors from the eastern states and Europe formed their own cattle companies,

and as the number of gold camps grew and more Indian lands were taken over, the capital invested and the numbers of cattle turned onto the open range increased sharply. In 1870, the agricultural census for Montana Territory tallied 2,000 head of cattle; by 1880 approximately 100,000 head of cattle grazed under Montana's big sky.

Completion of the Northern Pacific Railroad in 1883 made possible the shipment of stock into and out of Montana. Thereafter, speculation was fueled by reports that the first herds shipped to market by rail fetched over eighty dollars a head in Chicago, and by articles like one in the *Breeders' Gazette* in 1883 relating how a five-dollar steer could graze on free grass for four or five years and then be sold at market for ten times the investment.[10] Increased demand by investors for young steers to turn onto the range naturally followed. Rather than fattening their steers, midwestern farmers were more than happy to sell them at inflated prices to a clamoring western market. Those same cattle, when fattened and grown, were then shipped back to Chicago or Omaha. Agricultural historian Harold Briggs noted in 1932 that the "eastern stock commonly known in the West as 'pilgrims' or 'barnyard cattle' were crowded into westbound stock trains or driven to the ranges in large numbers. From 1882 to 1884 there were probably as many cattle shipped west as were received in the East."[11]

Large profits enticed stockmen to go to great lengths to get their beef to Montana grazing lands. With the end of the Civil War, thousands of surplus cattle populated the Texas ranges, and stockmen there looked for new markets and free grass. Over the decades that followed, the XIT Cattle Company became legendary for its drives of Texas longhorn steers. From the Texas Panhandle to the northern range was a journey of over a thousand miles that required months on the trail. The XIT blazed what became known as the Montana Trail, moving between twenty thousand and thirty thousand head of cattle into the territory during the late nineteenth century.[12]

Sheep raising took off in earnest in 1869 when John F. Bishop and Richard Reynolds herded fifteen hundred rams and ewes from Oregon into the Beaverhead Valley. The Beaverhead was the center of sheep raising in the territory for the next decade, and the flocks grew ever larger. Soon, enterprising sheepmen began moving large bands into the valleys of central Montana in flocks that numbered in the thousands. Wool was first freighted out of Montana Territory by ox team and then steamboat. But after 1883 and the completion of the Northern Pacific Railroad, Billings became a primary western shipping point, averaging 15 million to 16 million pounds of wool sent annually to Boston, Chicago, and all points east.

The number of sheep in the territory climbed steadily and by 1886 approached one million.[13] While competition for control over ranges and good, reliable water led to bitter enmity

The monitor-roofed horse barn (right) at Fort Custer was part of an extensive livestock facility at the military post established in 1877 near the confluence of the Bighorn and Little Bighorn rivers. The buildings are no longer standing.

The curving pediment of the brick mule barn at Fort Missoula presents an impressive façade. Shown at right in 1958, the barn is today one of the best remaining military barns in Montana.

between stockmen and wool growers elsewhere, Montana remained relatively peaceful, with many of the bigger operators running both cattle and sheep.

By the mid-1880s, the numbers of livestock grazing under Montana's big sky had grown exponentially, and every month saw announcements in local newspapers of more animals arriving. Some open-range outfits ran as many as thirty-five thousand head of cattle, a number reflecting the fact that ranchers had yet to understand the carrying capacity of western lands. In 1886, with the range overstocked, disaster struck. Cattle were stressed by poor grass conditions, and when winter set in and there was no shelter to block the raging winds and subzero temperatures, they died in droves. They died in coulees, and they froze standing up. The numbers were staggering. Many big cattle outfits lost up to 50 percent of their stock, while a few eastern Montana operations lost 90 percent of their herds. Together, Charlie Russell's famous painting of an emaciated steer—*Waiting for a Chinook, or Last of 5,000*—and stories of dead cattle snagged high up in trees that had drifted over with snow paint a harsh picture of the conditions that winter.

When the chinook winds blew in and the snows finally melted, the headcount was more than sobering; as many as 360,000 head of cattle died that winter.[14] Dead cattle washed downstream in rivers and creeks for months. The cattle boom went bust, and many corporate investors pulled out. Some operations regrouped, and plainsmen began to change their practices. Within a decade, the livestock industry recovered, and by the turn of the century, stock numbers were once again up—to over 1.5 million head of cattle in 1898 and to 6 million sheep in 1901, making Montana the nation's top sheep producer.[15] This time, however, the handling of livestock was different; ranchers more actively managed their herds. Moving livestock into sheltered valleys in winter, sending more to market in the fall, and growing and cutting hay for winter feedings became key to successful ranching.

The changes in ranching practices gave rise to additional barn construction—barns for horses, loafing sheds and other shelters for cattle, and hay barns for feed storage. Although cattle were still largely turned out to fend for themselves, cattlemen learned to put up tons of hay to carry their animals over the winter. As horses were more valuable and not quite so hardy, the earliest barns were commonly stallion barns or horse stables, often situated close to the ranch house in order to keep watch over the animals and guard against horse thieves. Until the era of mechanized transportation and farming, horsepower was the moving force behind successful agriculture, and barns to house teams of workhorses were essential.

Frontiers and Government

One of the biggest early markets for western ranchers was the U.S. government, and enterprising businessmen such as T. C. Power and I. G. Baker established trading empires on lucrative government contracts that included selling livestock provisions and beef and other food supplies to the U.S. Army and the Office of Indian Affairs.

Raising vast herds of horses for soldiering was critical in the era of animal power: horses were used for moving troops, engaging in active warfare, and conducting the heavy business of transportation, construction, and provisioning of military posts. The army required healthy horses and mules for its activities and built facilities to shelter them, such as the horse barn at Fort Custer, the brick mule barn at Fort Missoula (opposite), and the spacious brick cavalry stable at Fort Assinniboine (see page 202).

The plains of eastern Montana proved a productive place to raise horses, and the country's largest remount station was established at Fort Keogh near Miles City in 1900. Area ranches kept Thoroughbred studs and bred tough, rugged horses used in conflicts around the globe, including the Boer War and World War I. Miles City hosted livestock auctions and soon became the largest horse market in the world as thousands of Montana horses were sold to the armies of the United States, Great Britain, France, and Italy.

Other government agencies also had need for livestock. In the 1930s, the U.S. Forest Service set up the Ninemile Remount Depot west of Missoula, where saddle horses and pack mules were bred. The horses from Ninemile were raised for backcountry and fire-fighting details and used throughout the northern Rocky Mountain region by rangers and foresters through the mid-twentieth century.[16]

Age of Industry

Industry in western Montana, particularly copper mining and smelting at Butte and Anaconda, and wholesale logging across western Montana created a need for draft animals. Horses and mules labored in the mines of Butte, riding the hoists underground where a career of pulling ore carts through dark, smoky mine shafts awaited them. These animals would spend months in the mines, so long as they were healthy and the mines were running. Above ground, muscular mules and horses were bred and raised far away from the dust and heat of the copper mines.

Over the mountains in the western Montana valleys, horse logging with Percherons and other draft breeds was a major industry. The Anaconda Copper Mining Company maintained horse logging camps at Potomac and Ninemile, providing the horsepower for logging the millions of timbered acres needed to build and fuel the industrial copper complex of Butte and Anaconda.

Hungry miners, their families, and the growing metropolises also created demand for basics for the table—milk, butter, and cheese. Large dairy operations opened near the city of Butte in the broad bowl of Elk Park, where buildings from these farms still stand today.

Fruits of the Plains

With the passage of the Homestead Act of 1862, the stage was set for the blossoming of Montana agriculture. The Homestead Act broke the lands of the West into private parcels, allowing citizens to own 160 acres if they could "prove up," that is, move onto the land they had chosen, build a structure on it measuring at least eight feet by ten feet, improve the land by planting crops, and live there for five years. The act laid out the pattern for cutting the vast blanket of Montana's virgin soils into a patchwork of land claims and forest tracts.

The opportunity to own land drew people who had little and those who wanted more. Between 1862 and 1900, people flowed to unclaimed federal land in the West from the eastern seaboard, the Midwest, and foreign nations. In Montana during this era, they homesteaded farm properties near mining towns in the western half of the state and settled on rangelands where sheep and cattle operations were launched in the central part of the state.

It was the Enlarged Homestead Act of 1909, along with recruitment by the transcontinental railroad companies, that threw open the floodgates to agricultural settlement across the full breadth of the state. In all, almost 2 million homestead claims were filed as people flocked to the Great Plains. Subscribing to the bootstrap thinking of such founding fathers as Benjamin Franklin—"diligence is the mother of good luck"—homesteaders attempted to draw a livelihood from the soil. Their stories are colorful and filled with hardship. Many lived in tar-paper shacks or crude dugouts in the ground; some were known to harness themselves to the plow to break soil. Far beyond settled towns, loneliness, howling winds and deep snow, and lack of supplies could drive those not toughened to the lifestyle into the depths of depression.

The transcontinental railroad companies depended on establishing towns and rural populations in the West to make the long haul across expansive western territories profitable.

(below) The Anaconda Copper Mining Company's horse ranch near Potomac in the Blackfoot Valley used strapping horses, like the skidding teams pictured below in 1910, to supply wood for the copper mines and smelters in Butte and Anaconda. One of the company's classic mountain horse barns is pictured above in 1911.

They promoted the territories as a breadbasket of America and embraced the dryland farming theories of Hardy Webster Campbell, whose philosophy of deep tilling the soils—and then mulching crops with the dust to help them hold moisture—became the gospel of dryland homesteaders. "The camel for the Sahara Desert, the Campbell Method for the American Desert" was the slogan Campbell spread in 1909 at the Dryland Farming Congress held in Billings as he assured farmers in the semiarid West that rain would indeed follow the plow.[17]

Coming from back east and places of greater rainfall, Montana homesteaders looked across the dusty High Plains and through starry eyes saw fields of golden wheat and fruit trees. They saw docile dairy cows with tinkling bells ambling across green pastures. They saw hens and turkeys, sheep and cattle, oats and winter wheat spilling over bushel baskets and filling their granaries. They saw gardens planted with rhubarb, huge cabbages, succulent beans, and red tomatoes. And for a while, at least during the unusually wet years of 1910 through 1915, the earth gave forth these things, thanks to generous rainfall and the natural richness of the prairie.

But irrigation was soon recognized as the key to expanding and stabilizing agriculture throughout Montana. With the Reclamation Act of 1902, irrigation projects were launched across the state, creating reservoirs, diverting water into canals, and establishing a complex canal-and-ditch system off every major waterway. Completed in 1907 just east of Billings, the Huntley Irrigation Project was the second federal Bureau of Reclamation irrigation project in the nation. Such projects brought thousands of marginal acres into greater production, leading to large-scale irrigated farming. Land was planted in orchards and crops that demanded more water, including sugar beets, which sparked a nascent sugar industry in the Missouri and Yellowstone valleys.

These were the years of plenty, and Montana must have seemed a proverbial land of milk and honey. Barns grew larger with the expansion of dairying near the state's growing towns, especially in central Montana, where the Judith Basin dominated state dairy production, and in the Gallatin Valley, where many second-generation barns reflected the growth of dairy operations.[18] Milk, cheese, and butter produced from Holsteins and Guernseys were sold to thousands of city dwellers. Bumper crops of wheat were harvested throughout the state, and products like Montana's Cream of the West were born.

Homestead acts passed in 1862 and 1909 promoted agricultural settlement on the Great Plains, where almost 2 million homesteads were claimed by homesteaders. This homesteader was photographed breaking eastern Montana's prairie sod circa 1910.

Bad luck, however, soon arrived, as years of plenty during the heyday of the homestead movement gave way to years of drought and hardships of biblical proportion. Skies filled with grasshoppers, and Mormon crickets ate the handles off farm tools. Dust blew so thick it blackened the sun, covered fences, and crept into every nook and crack in homestead homes and outbuildings.

The homesteaders struggled to survive. In the end, the challenges overpowered many of them, especially those who did not have access to irrigation. Between 1919 and 1925, 2 million acres in Montana went out of production, half the farmers in Montana lost their land, and a mass exodus ensued.[19] Among those who were forced out, many just walked away, leaving dishes on the table, turning their animals out to forage for themselves, and forfeiting their land to the bank.

In the end, some sixty thousand Montanans left the state, about 10 percent of the peak population during the boom.[20] But farmers and ranchers with water rights survived the drought, and through the 1920s and 1930s, they expanded their range by buying up failed homesteads. Farms diversified, larger-scale ranches emerged, and finally the hard lesson about the carrying capacity of the land was learned. The key to success for a farmer or rancher was larger acreages, access to reliable water, and more carefully managed herds.

On the rebound, some stock growers welcomed paying guests to their ranches. The phenomenon of "dude ranching" began in the 1880s, and its heyday lasted almost eighty years. To many families, the extra funds made all the difference

in getting back on top. Dick Randall, a.k.a. the "Father of Montana Dude Ranching," established the OTO Ranch as Montana's first dude ranch, offering a rider's mountain paradise on over three thousand gorgeous alpine acres just north of Yellowstone National Park.

To the east, just south of the Montana line, Sheridan, Wyoming, was a national center of the industry, and again the railroads played a role. For years, a train called the "Dude Rancher Special" transported easterners to Sheridan to take in the western life. Much like the fly fishermen of the twenty-first century, eastern adventurers, in starchy new Levis and spanking white Stetsons, stepped off the train ready for a colorful western vacation. Trail rides, cow punching, branding, and roping kept the dudes busy. Many found an honesty on the working ranch that kept them coming back year after year; a few found romance, married, and never went home. With their rustic lodges, fine barns, and cozy cabins, the OTO, the Quarter Circle U, and the Bones Brothers Ranch (see page 266) were among Montana's finest dude ranches.

As agriculture moved into the mid-twentieth century, western ranchers and farmers tied into a growing national market. White-faced Herefords, valued for their large size, thick hides, and ability to survive harsh conditions on the range, became the symbol of the cowboy West. With time, these attributes were less appreciated than the marbling and genetics that other breeds brought to the mix, and today the Black Angus is by far the dominant breed. Still, when you look into the white mask worn by many Angus steers, it is the face of the Hereford—crossbred with the Angus to help them adapt to the High Plains—that looks back at you.

Participating in roping and branding were among the roundup experiences shared by paying guests during the dude ranch era, as photographed in 1939 on the 3 Circle Ranch (left) and the Quarter Circle U Ranch (right), both in the Tongue River Valley in Powder River County.

CHAPTER 3

Barns of the Western Valleys

The Bitterroot and the Clark's Fork Drainages

In the fall of 1841, Father Pierre-Jean De Smet left St. Mary's Mission for Fort Colville on the Columbia River, an outpost of the Hudson's Bay Company. There he bought some cattle, a few horses, and the first seeds to get farming under way at the mission. When the spring came, the Jesuits and the Salish planted large gardens, and the fertile valley floor soon yielded carrots, onions, beans, and wheat. In 1845, Father Anthony Ravalli added a flour mill; the following year, De Smet transported "ploughs, spades, pickaxes, scythes and carpenter's implements" up from the Columbia River by horse and wagon to St. Mary's and to a new mission near Fort Colville.[1]

These supplies gave the Jesuit fathers the tools to build, and build they did. By 1846, St. Mary's included "twelve houses, neatly built of logs, a church, a saw mill, and other buildings for farm use."[2] With seven thousand bushels of wheat and more than four thousand bushels of potatoes, the farm storage buildings overflowed with the fruits of the labors of the missionaries and the Salish. Although none of the early buildings remain, the mission's farm buildings appear to have been the first *bereaerns* in the history of Montana.

In 1850, attacks by the Blackfeet, who stole horses and shot cattle—and other factors as well—caused St. Mary's to close. The property was sold to Major John Owen, who opened a trading fort there. The former mission became a way station to settlement, drawing hopeful farmers, retired soldiers, and trappers to the valley. Governor Isaac I. Stevens's railroad survey expedition wintered over there in 1853–1854, establishing a cantonment near Fort Owen and a market for provisioning exploratory parties.

With more than a dozen fledgling farms and a small grist mill nearby, John Owen was decidedly upbeat as he noted in his 1860 journal that "crops are flourishing" and that "farmers need have no fears for their wheat, it is made. The grass is better than I ever saw it before in this country." He also recorded the building of his fort, and his daily entries like "making adobes, hauling rock, cutting timbers" resonate still today, recalling the rhythms of early-day settlement as experienced hands were tasked with building everything from forts and roads to homes and barns.[3]

That same year, 1860, the Mullan Road connected Fort Benton on the Missouri River and Walla Walla in Washington

(opposite) Porter Barn, Flathead County, built 1908

The prized racehorses of Butte copper king Marcus Daly enjoyed luxurious quarters in Tammany's Castle, the horse barn on his stock farm outside Hamilton in the Bitterroot Valley. The Castle, right, is pictured in 1899, and in the photograph below is Tammany himself (in doorway) and his stablemate Hamburg.

Territory, opening an important route into western Montana and its newly discovered goldfields. The Homestead Act of 1862 legitimized land claims, and farmers edged deeper into the rich bottomlands of the Bitterroot Valley. The town of Stevensville was born, and Bitterroot "squash peddlers" began hauling crops and beef cattle to the Grasshopper Creek diggings at Bannack and the mines at Virginia City. And even though early farming proved almost as risky as mining, farmers gamely battled grasshoppers and bad weather as demand for fresh vegetables made the payoff high.

To the north, the Bitterroot and four other valleys merge to form the broad Missoula Valley. As Meriwether Lewis called it in 1806, the "vally of Clarks river" was "extensive beatifull level plains and praries. the tops of the hills and mountains on either hand are covered with long leafed pine larch and fir; near the river the bottoms are timbered with long leafed pine and cottonwood."[4]

Métis French Canadian traders Louis Brown, Moise Reeves, and Baptiste Ducharme arrived here between 1857 and 1859, leaving the fur trade to farm the river bottom and take advantage of that abundant forest to build log homes and barns. A decade later, Thomas Meagher noted some fifty men at Frenchtown and described excellent orchards of "not less than three hundred fruit trees—pear, plum, peach and apple" and farms ranging from "20–100 acres under the choicest and ripest cultivation."[5] The settlement became a permanent village along the Mullan Road, and the Missoula Valley grew as a trading and agricultural center, its mild climate conducive for growing all kinds of crops. Meanwhile, in 1860, C. P. Higgins and F. L. Worden had established a trading post across the valley at Hellgate, and over the years, the city of Missoula grew between these two outposts.

At Frenchtown, in this prerailroad era, log building continued as the order of the day. The remaining barns include some of Montana's earliest and finest, reflecting a well-honed ethnic building tradition, carefully fitted and highly distinctive. Among the more than a dozen barns and granaries that survive are some exceptional cribbed log buildings: two double-crib granaries and two English-style barns with gable roofs, square-hewn log walls, and expertly notched corners as well as round log barns for horses and hay. A twentieth-century gambrel-roof horse barn marks a later era in the community's history.

In 1886, industrialist Marcus Daly moved into the Bitterroot Valley to establish the company town of Hamilton as a timber satellite of the growing Anaconda Copper Mining Company. In contrast to the small-scale and hand-crafted nature of the frontier buildings at Frenchtown, Daly's town transformed the valley, linking lumbering and farming enterprises to the industrial mining empire at Butte and Anaconda. His was an industrialized approach of the highest order.

On the outskirts of Hamilton, Daly built the Bitter Root Stock Farm on a scale that befitted a copper-mining magnate. His was an enormous twenty-two-thousand-acre farm with blooming orchards, golden grain fields, acres of sun-drenched vegetables, fattened cows, and glossy horses. Daly's true passion was his stable of racehorses—fleet, blooded horses with lyrical names like Lady Peel, Montana, Killarney, and Moondyne. He modeled the stock farm after Irish and Kentucky estates and built substantial barns, training facilities, and pastures to service nine hundred horses. For his favorite racehorse, Tammany, Daly built what became known as Tammany's Castle, a brick stud barn with round-arched doorways, and employed the very best veterinarians, trainers, and stablemen around (see page 58).

Even in these moist valleys west of the Continental Divide, annual rainfall averages just twelve to sixteen inches, well below the twenty inches considered necessary for nonirrigated farming. The Jesuits were the first to irrigate in Montana, and later settlers, including Marcus Daly, built a Bitterroot Valley irrigation system. This culminated in the eighty-mile-long "Big Ditch," completed after Daly's death in 1910. Along its route, twenty-five thousand acres were subdivided into orchard tracts, sparking an apple boom that lured easterners to a "Land of Opportunity" where apples, pears, cherries, and plums grew in abundance. In the Missoula Valley, the 1930s Frenchtown irrigation project brought diversified farming of such crops as sugar beets, potatoes, and peas and increased alfalfa hay production and forage, in turn tripling the capacity for raising cattle in the area. The barns of the twentieth century reflect the expanded prosperity this brought to farming communities.

The Flathead Drainage

In March 1812, Canadian explorer David Thompson overlooked Flathead Lake and declared it a "fine sheet of water." Four decades later, in 1854, Father Adrian Hoecken founded a new St. Ignatius Mission near Flathead Lake, where, within a year, "our good brother McGean had cut 18,000 rails and placed under cultivation a large field." The missionaries introduced cattle and provided plows, tools, and seeds "gratis to everybody," and farming was launched in the Mission Valley.[6] The building of a mission to succeed St. Mary's got under way, and all buildings were made of log, from the chapel and cabins to the blacksmith's and carpenter's shops. Within a few years, the Jesuits had built water-powered flour mills, sawmills, and a large, timber-framed barn.

Father Adrian Hoecken relocated the St. Ignatius Mission from Washington Territory to the Flathead Valley in 1854. Within a few years, the Jesuits had built water-powered flour mills and sawmills and a large timber-framed barn. In this 1867 illustration, the barn features a broad gable roof and a side shed, and it appears to be quite sizeable.

By the late 1850s, the Salish owned a herd of over four thousand horses, and St. Ignatius Mission had a thousand head of cattle and Montana's first flock of sheep.[7] A reported fifty small Native farms took hold, and St. Ignatius Mission produced a surplus of flour, which was sold to the American Fur Trading Company.

Early settlers north of St. Ignatius discovered a microclimate around Flathead Lake, where moist breezes gather over the water and gently kiss the tops of trees and fields as they move across the nearby land. Outside events catalyzed settlement and farming in the region. The discovery of gold in British Columbia in the 1860s created a travel route north from Missoula, and completion of the Northern Pacific Railroad through Ravalli in 1883 brought an increase in the region's population. By 1892, when the Great Northern Railway laid tracks through to Whitefish and beyond, there were already burgeoning timber and farming industries in the area. The planting of seven hundred cherry trees on the west shore of the lake in 1929, and several hundred more trees the following year, gave rise to the Flathead cherry industry.

Barns in the Flathead region evolved with changes in the valley. By the early twentieth century, the timber industry, sawmills, and railroads made commercial building materials more readily available, and many large structures were erected to serve agriculture in the valley—storage barns for hay, an enormous cherry warehouse, and livestock barns to hold dairy cows and draft horses for timber work. In this era of animal power, Frank Porter, owner of a local lumber company, constructed a barn overlooking the lake that was large enough to house four teams of workhorses. When the Porter Barn

Duncan McDonald owned one of the earliest farms on the Flathead Reservation, and his hand-hewn log barn, built circa 1910, was one of the first barns on the reservation.

was completed in 1908, the local paper proclaimed the timber-framed structure "the largest in the entire valley" (see page 42).

To the west and north, glacier-fed rivers like the Flathead and the Kootenai carve their way through a deeply forested region that was opened to the fur trade in 1809 when David Thompson put up the Saleesh House along the Clark's Fork River near present-day Thompson Falls. Gold strikes in the 1860s drew miners to this remote region, but it was not until after the coming of the Great Northern Railway in 1892 that lumbering became profitable. At the same time, the population grew, as did the number of farms along the Kootenai River.

Still, the number of people who lived in Montana's most northern woods remained comparatively few, and hunting, trapping, and subsistence farming continued to be the way of life. In contrast to the topography of the broad lower Flathead Valley, valley floors in the upper Flathead drainage are narrow, and shadowed forests cloak the steep slopes surrounding them. This is Montana's snowbelt, where valleys are blanketed in snow many months of the year. Reflecting limited budgets and Nordic conditions, barns in northern Flathead, Glacier, and Lincoln counties exhibit a sturdy regionalism—many are mountain horse barns with thick log walls, steep gable or gambrel roofs to shed heavy snow loads, and room to house essential livestock and to store winter feed in the loft.

Public lands and federal projects played a major role in the economy of this region. The creation of national forests and of Glacier National Park in 1910 altered the trajectory of life in the woods. National forests provided a setting for those who drew their livelihood from the land, and major companies established lumbering camps to provide for the building of western Montana towns. Expanded hydroelectric facilities provided reliable irrigation that broadened farming in the decades to follow, and the number of farms in the region proliferated in the twentieth century.

The barns in Montana's western valleys today reflect this history, marking the transition from the very earliest farming to the years when mining, lumber, and agriculture ranked one, two, and three in the economy of the state.

St. Ignatius Barn LAKE COUNTY, 1941

Among the first buildings put up in the Flathead Valley were the barns at the St. Ignatius Mission—barns built under the direction of Father Hoecken by Brother Francis and Brother McGean after their arrival in 1854. Father Pierre-Jean De Smet mentions the brothers in his writings about the achievements of the mission: "They both worked hard in bringing the mission to its present state of perfection, building successively a windmill, blacksmith's and carpenter's shops, barns, cowsheds, etc., besides an excellent chapel, in addition to a large dwelling-house, of hewn timber, for the missionaries."

A photograph dating to the early 1880s depicts the major buildings at the mission, including a flour mill, a blacksmith shop, a bakery, dormitories, and a barn. The log barn, composed of a gabled center section and attached shed-roof wings, was one of the largest structures of the complex. More barns would come in later years to accommodate a variety of functions. Unfortunately, none of these barns exists today.

In contrast to the rough-hewn early buildings at the mission, the handsome red cattle barn now standing near the old St. Ignatius flour mill was built in 1941, one of the last buildings constructed as part of the mission. Although the milling business had been a sure source of income for the mission since its establishment, the mill was abandoned in 1935 in favor of developing a beef cattle and dairy operation. The Jesuits raised Hereford cattle and also ran a successful dairy for more than twenty-five years before selling to a private ranch company in 1966.

The 1941 cattle barn (opposite). Construction on the St. Ignatius Mission church (above) began in 1891.

The large log barn was either demolished or perhaps simply recycled into the barn existing today. The lower floor of the barn as it is today is constructed of large, hand-hewn posts and beams, a method of building surely used in the early mission buildings but one that was long abandoned by 1941 in favor of lighter frame construction using milled lumber. Upstairs, the thoroughly modern second-story hayloft is supported by laminated arched trusses, which create a large unobstructed storage space.

The flour mill is one of the few buildings from the 1880s still standing at St. Ignatius.

Some of the timbers in the 1941 barn may have been recycled from the original buildings of the mission.

Somers Barn

FLATHEAD COUNTY, 1903

Missoula architect A. J. Gibson designed this stylish carriage barn to accompany Somers Lumber Company president John O'Brien's elegant mansion overlooking the town of Somers. Gibson was the designer of the Missoula County Courthouse, Main Hall at the University of Montana, and many homes in Missoula. The O'Brien carriage barn, with its clipped gable roof, neat white trim boards, and fine row of carriage doors, is a good example of the level of style achieved by leading architects who branched out to design agricultural buildings. The barn held a row of stalls for horses and space for a buggy or two. Hay was stored in the loft, and the rear and basement rooms provided a residence for a hired hand or driver. The barn was heated, as were the mansion and a few other houses in Somers, by steam pipes running up from a heating plant at the mill.

O'Brien lived at the lake for only three years before moving to the Pacific Coast to develop timber operations there. From 1931 to 2005, the Somers mansion and barn were owned and well kept by the McDevitt family, a tradition continued today by the C. D. Didier family.

Somers Lumber Company Barn

FLATHEAD COUNTY, CIRCA 1920

This expansive horse barn was built by the Somers Lumber Company in the 1920s to house draft horses at a time when animal power was essential for a variety of tasks all around town. Built with heavy timber posts and braces as well as two-inch-thick by twelve-inch-wide floorboards, the barn was sturdy and well suited to shoulder the weight of its huge inhabitants.

Somers was a company town and the center of the timber industry in the upper Flathead Valley. In 1901, John O'Brien established the O'Brien Lumber Company to supply lumber to the Great Northern Railway. The company, later called the Somers Lumber Company, used Flathead Lake as the holding area for the freshly cut logs and established a town of 120 dwellings for its workers. Up above the town, O'Brien built a handsome mansion and carriage barn for his family (opposite).

The Somers Lumber Company Barn housed workhorses until it was dismantled in 2010.

Olsen Barn

FLATHEAD COUNTY, 1910

The Olsen Farm is a joy to see as working farms become increasingly rare in the Flathead Valley. The present owner, Thomas Nonemacher, still uses this 1910 dairy barn, built originally for widow Bertha Olsen, to house his own small flock of chickens, a dairy cow, a few Scottish Highland cattle, and some pigs.

This sturdy, multipurpose building also illustrates twentieth-century balloon-framing methods. The hayloft, with its hood and hayrack, was designed with storage and ventilation in mind. The upper wall of the loft is partially open, allowing for more air circulation. Nonemacher uses a conveyor system to carry hay bales up into the loft.

Scottish Highland cattle

Hay still fills the loft of Bertha Olsen's 1910 barn.

Loafing shed

Original homestead house, 1893

McCarthy-Pederson Barns

FLATHEAD COUNTY, 1893, 1905

The modest farm buildings on the McCarthy-Pederson Farm reflect the early days of agriculture in the northern Flathead Valley. John McCarthy was one of the first permanent settlers in the area south of Creston. He was an enterprising farmer who took advantage of the fertile ground to raise crops for the fledgling railroad community of Kalispell. McCarthy bought the land in 1892 and began building a small house and granary. When the Pederson family acquired the farm in 1907, they set about improving the farm with a larger house and a loafing shed for their cattle. While the Flathead Valley boasts many large and impressive barns and outbuildings, with grand histories to match, these structures on the McCarthy-Pederson Farm reflect the often-overlooked history of the many small diversified farming operations in the region.

(left to right) Calving barn, 1893 granary, and loafing shed

Snell Barn

FLATHEAD COUNTY, CIRCA 1900

The Snell Barn is an excellent example of an English or three-bay barn with heavy timber framing. Its massive timbers are mortised, or pegged together, without nails, providing more than adequate support for the roof and hayloft floor. Inside, a cross-shaped floor plan allows for a wide alley in the center bisected by a narrower alley, with stalls opening off of it.

Over time, as agricultural pursuits expanded and adapted to regional climates in America, barn design evolved to accommodate a variety of livestock and grain-storage needs. Most ranchers in the West oriented their barns on the long axis, locating the main entrance in the gable end, which provided room for a few more stalls. Also, as barn building technology advanced and the availability of milled lumber increased, the method of timber-frame construction was replaced by the more contemporary stick-framing, or balloon-framing, method.

This early-twentieth-century barn is something of a throwback, built using traditional building methods at a time when nearby barns give evidence of employing a lighter, simpler framing method. It is interesting to note that the September 20, 1907, issue of the local newspaper, *The Daily Interlake*, advertised a very similar, though larger, barn plan with a gable roof and central alleyway. The article may have influenced this barn plan, while the late use of timber framing, here and elsewhere in the valley, may have been due to the ready availability of so much local timber and so many skilled log workers.

Hand-hewn timbers fastened with pegs support the Snell Barn roof.

Porter Barn FLATHEAD COUNTY, 1908

Frank Porter, a well-known Flathead Valley pioneer and timber man, had this giant horse barn built as a reflection of his wealth and prosperity. It is now a reminder of the early settlers who made their living in the timber industry and farming. The barn is ninety-five feet wide with a central drive-through alley. The barn held sixteen workhorses and had separate tack and grain rooms as well as a large open area for shoeing and horse care. The hayloft was divided by a large central threshing floor directly above the alley and was flanked on both sides by cavernous hay-storage space. It was constructed using heavy timber framing, despite the fact that balloon-frame buildings were becoming the norm.

The Kalispell newspaper, *The Daily Interlake*, followed the construction of the barn closely and may have influenced its design. The paper ran several articles on barn building in the early years of development in the valley. One article in September 1907 promoted a design very similar to the Porter Barn, with a drive-through alley and stalls for sixteen horses. Although the design in the paper had a gable roof, instead of a gambrel roof, and was smaller, it may explain the use of the central-alley format. Another influence may have been Porter's heritage in Wisconsin, where this design was quite common.

Other articles in the paper reported that Porter hauled the lumber to the site in June, a herculean task considering the girth and length of some of the barn's timbers and that most builders completed this task in winter when the ice made the haul much easier for the horses. A later article reported that Porter's barn was to be "the largest in the entire valley." And when the barn was finished in late August, the hoedown party held in it was given special coverage. People from all over the valley attended, and there were two bands for entertainment.

It was also reported that the Lower Valley Stars baseball team hosted the event as a fund-raiser for the team.

Porter bought the 160-acre ranch property in 1898 and would later buy up other properties in the area to finance his many business interests. He held the ranch until 1918, longer than any of his other properties, when he sold it for one dollar in exchange for over fourteen hundred acres held by the buyer, Charles A. Lynch, one of the wealtheir cattlemen in the valley.

Mellem Barn

FLATHEAD COUNTY, 1913

The Mellem Barn, along U.S. Highway 93 near Kalispell, is a well-known landmark in the area. The date 1913 is painted on the gable end to mark the year this barn was completed by father and son Ole and Emil Mellem. It was probably the first major carpentry project undertaken by Emil, who, at age nineteen, was still learning the trade. His daughter Shirley recalls that Emil spent a week hand-hewing the central support beam for the roof. She also remembers her father boasting that he installed the weathervane (now gone) on the roof peak without wearing a harness or rope to guard against falling. Emil Mellem went on to build many beautiful log homes in the valley.

Anderson Barn

GRANITE COUNTY, LATE 1920s

This log barn is located near the small settlement of Jens in the beautiful Deer Lodge Valley. Most likely built by the Anderson family, it is a good example of a mountain horse barn built in the late 1920s. A single-crib log structure, its entrance is centered under the gable. It has a central passage that gives access to stalls and/or storage rooms and hayloft. This barn type proliferated in the late 1800s in southwest Montana with the beginning of stock raising in the 1860s, and it is interesting to note that it continued to be used well into the 1930s.

Although a mountain horse barn built in the pioneer days of the 1860s might look similar to one built in the 1920s, a few interior construction details can help to reveal when a barn was built. Older barns built in Montana between 1864 and approximately 1890 were often framed with heavy timbers or large posts and beams mortised together with wooden pegs. In general, after 1890, advances in building technology and the availability of milled lumber and manufactured nails allowed ranchers to build barns with much-smaller lumber or logs that could be nailed together. The Anderson Barn is easily identified as a post-1890 barn because it employs long nails spiking the exterior logs at the corners and nailed, small-dimension logs in the rafter system.

Glacier National Park Barns

GLACIER AND FLATHEAD COUNTIES, 1926–1935

Lubec Ranger Station Barn

The Lubec, St. Mary, and Walton ranger station barns are all fine examples of the horse and mule barns built in Glacier National Park in the 1920s and 1930s. Following passage of the Glacier National Park enabling act in 1910, the U.S. Department of the Interior found itself responsible for the development and administration of 1 million isolated and rugged acres. Park managers faced many unprecedented problems, not the least of which was to design and fund a system of facilities that would serve the visiting public and preserve and protect the area's scenic and natural values.

The Lubec Ranger Station Barn was built in 1926, using a typical rustic design in keeping with the park's architectural philosophy of employing "pioneer log construction" for

(above) Walton Ranger Station Barn. (left) The wooden catwalk of the St. Mary Utility Area Station Barn allows transport of hay from the loft to where it can be thrown into the feeding troughs below.

buildings. Although the Lubec Station was dismantled in 1976, the barn was carefully moved and reconstructed at the 1913 St. Mary Ranger Station that year.

The St. Mary Utility Area Station Barn (1935) and the Walton Ranger Station Barn (1937) are two of many buildings constructed by the Civilian Conservation Corp (CCC) during a substantial building campaign. According to Albert Good, an architectural consultant to the National Park Service in the 1930s, the rustic style was one that "through the use of native materials in proper scale, and through the avoidance of severely straight lines and over sophistication, gives the feeling of having been executed by pioneer craftsmen with limited hand tools. It thus achieves sympathy with natural surroundings and with the past." Today, these humble barns, along with many other rustic structures, represent the successful efforts of the National Park Service to integrate necessary buildings into this wilderness landscape.

Goats congregate on the ramp where wagons once entered the barn.

River View Dairy Barn

GLACIER COUNTY, 1917

The River View Dairy was established in 1903 by Romain Chasse in the early days of homesteading in the Cut Bank area. Chasse had been in Montana since 1878, raising cattle at different locations for the Jesuits. After moving around for several years, he found choice farmland right alongside Cut Bank Creek near the railroad bridge. Chasse, his wife, and fifteen children started a truck farm and dairy here, selling vegetables, milk, and ice in the budding town of Cut Bank. George Chasse, who was twelve at the time, was the delivery boy and remembered that the ladies at the bordellos left him the best tips.

The Chasses set about building their barn soon after arriving, situating the forty-eight-cow "bank" barn on the upslope of a hill, or bank, making it possible for wagons to access the hay doors directly from the farm's driveway. Taking the design a step further, Chasse built tall, wide hay doors on both ends of the barn so that a wagon could drive into the barn and continue through the large central alley and down the hillside. Upon completion of the barn, the local newspaper ran an article praising the innovative design.

Bank barns like this one are less common in Montana and are often associated with more prosperous farms. Chasse's farm was indeed prosperous, and the farm likely financed construction of the family's three-story Cut Bank Hotel. In 1917, Chasse moved into town to run the hotel full-time and sold the farm to Frank Corrigeux. Corrigeux and his family ran the dairy farm until 1940 and the ice business until 1948. They sold the dairy in 1969, and today the barn is used for milking goats and for lambing.

Little Beaver Stock and Poultry Farm Barn

SANDERS COUNTY, 1908

This century-old hay barn was one of two barns built in 1908 by homesteader Edward J. Thompson. The other barn, which was identical to this one, was in a field nearby and was destroyed by forest fires that swept the area in 1910. (Note the lack of trees on the hillside in the historic photo on the facing page.) The horse stable was housed under the shed roof portion on the left, and a pulley in the hayloft came out with a rope-and-grapple assembly to pick up the loose hay. The barn was built to hold up to sixty tons of loose hay.

The Little Beaver Stock and Poultry Farm is a good example of the community effort that went into early-day farming. The barn was built with assistance from Thompson's neighbor Ernest Butte, who helped haul the timber by team and wagon from the Hartman place, a mile and a half away. And, at harvest time, as seen in the historic photo, a large crew, including women, gathered from all around to cut and bale grass into hay. After one farmer's fields were cut, neighbors would move on to the next farm to get everyone's hay put up for the winter.

The Meadows family, the barn's current owners, celebrated the barn's one hundredth birthday in 2008, inviting locals to tour the barn, reminisce, and take part in a group photo to match the photo from "way back when."

Owners Bill and Helena Meadows celebrated the one hundredth birthday of their hay barn with a new photo (above) to match the historic one, which was taken by a traveling professional photographer (below).

Boyer Barn

MISSOULA COUNTY, CIRCA 1865

This exceptional double-crib log granary near Frenchtown was built in the mid- to late 1860s on the Louis Brown Ranch. It is among the oldest barns in Montana. The plan of the log granary features an alley at the center to allow for threshing or storage and single-room "cribs" on each side for grain stockpiling.

Traditionally, the central drive functioned as a threshing floor. Grain could be beaten or flailed to break down the stalks, and a good breeze would then blow away the chaff to winnow the crop, leaving the cleaned grain ready for storage.

The massive, hand-hewn square logs and the fastidious dovetail notching on this building suggest that its builder was quite skilled at his trade. This type of granary, which stands here in its original condition, is a rare survivor of its type in Montana, as many log buildings like this one were torn down, altered, or moved.

A logger and gold miner, Joseph Boyer of Quebec worked for Louis Brown, perhaps helping to build his log barns. Boyer eventually bought 621 acres, including the original log house and outbuildings, from Brown in 1896. Since then, the ranch has remained in the Boyer family. Today, Joe Boyer Jr. still uses the barn as a granary on the ranch, where he raises wheat, barley, oats, hay, hogs, and cattle.

(left) Double-crib granaries, featuring a drive-through alley for threshing, are rare in Montana.

The two barns on the Scheffer property represent two very different eras of barn building. The log barn dates from 1880 and the red barn from 1924.

Scheffer Barns

MISSOULA COUNTY, 1880, 1924

Tom Scheffer and his daughter Gyla

Tom Scheffer was only six years old in 1924 when his father, Ralph, and relatives built the big red horse barn on the family ranch. By that time, the ranch had been in the family more than sixty years—since the mid-1860s when Ralph's grandfather, Eusebe Scheffer, a French Canadian, located near Frenchtown and began raising hogs to sell to loggers in the nearby forest camps.

Eusebe Scheffer built the log barn in 1880, a simple yet extremely well-built gable-roofed building and likely no different from many he had grown up with or seen in his native Canada. Each log was hand-hewn into a perfect square, then stacked so tightly that no chinking was necessary. The barn was probably a multipurpose barn used for workhorses, cows, tack, and grain storage. In Tom Scheffer's day, it was used as the stallion barn and then later converted for use as a milk house. The Scheffers raised Percheron horses for the Anaconda Company's Ninemile logging camp and pack mules for the Ninemile U.S. Forest Service Station. "Horses is what they used for power; that's all there was," said Tom Scheffer, as he remembered that the family did not have a tractor on the property until 1935.

In contrast to the traditional barn of the 1880s, the twentieth-century horse barn was designed by an Iowa barn-plan company. Typical of the era, it features a Gothic arch roof and was built with smaller lumber cut on the family's ranch and planed on-site at their sawmill. The upper portion originally housed loose hay, and the lower half was home to draft horses. Hay chutes were used to send loose hay down to the horses in their stalls. Later, the barn was used for dairy cows.

The barns on the Scheffer Ranch near Huson represent two distinct types of Montana barns, and they are the legacy of seven generations of Scheffers on the property.

William Ford Ranch Barns

RAVALLI COUNTY, 1914–1917

Although the three very well-preserved barns on the historic William Ford Ranch are excellent examples of an early-twentieth-century model dairy operation, they housed dairy cows for only a few short years, from 1914 to 1917. William Ford and Judge Howard K. Hollister, wealthy Ohio businessmen, purchased the Shelton Ranch in the Bitterroot Valley in 1914 at the insistence of their sons, who were wooed to the West by land promoters, the breathtaking scenery, and the promise of a quieter life away from the city.

Ford and Hollister planted thirty-five hundred apple trees (a popular crop at the time), brought in a herd of one hundred purebred Holstein cows from Ohio, and constructed a large house and three large barns to support a model dairy operation. The three matching gambrel-roof barns were handsomely finished with clapboard siding, decorative door trim, cornice returns on the eaves, and twin ventilators, with cow weathervanes on each roof. One barn was for milking, one for feeding, and the third for workhorses. After a year, however, the two sons realized that farming was hard and sometimes lonely work, and they returned to their familiar city lives.

After the death of Hollister, William Ford continued to manage the ranch from a distance, though he eventually moved to the property with his second wife and two young daughters. Despite having the best buildings and dairying equipment, the operation was not profitable, and a few years later, Ford replaced the Holsteins with Hereford beef cattle. When Ford died in 1935, his wife and daughters continued to spend summers in the valley, and even ran the operation as a guest ranch for a few years. The ranch was sold after May Ford's death in the 1950s. The name was changed to Chief Joseph Ranch in the 1960s, and the Shupp family once again opened it to guests in 2004.

Tammany's Castle

RAVALLI COUNTY, 1894

At the time of its construction, Tammany's Castle was undoubtedly the most luxurious horse barn in Montana. While most Montana horses in 1894 would have been lucky to reside in an unchinked log stable with a leaky roof, for his favorite racehorse, Tammany, copper magnate Marcus Daly built a two-story brick castle, complete with Romanesque arches, colonial-style cupolas, and oak paneling in the stalls. This lavish stable was part of the Bitter Root Stock Farm, Daly's gentleman's estate just outside Hamilton.

Having made his fortune in mining and smelting in the late 1860s and 1870s, Daly arrived in the Bitterroot in 1886 to pursue new interests in stock farming and timbering. He purchased and remodeled the homestead of Anthony Chaffin, began acquiring surrounding farmland, and bought timber properties in the area. By 1889, his racehorses were winning on eastern tracks, and he had hired a resident veterinarian at the stock farm, a luxury unknown in the West. Daly purchased Tammany in 1890, and when the horse won the New Jersey Suburban Stakes in 1894, the first large purse won by the Stock Farm, Daly built Tammany the horse barn of all horse barns, complete with cork flooring and velvet lining in his stall. Daly retired Tammany at the end of 1894, but the horse remained an esteemed resident of the Stock Farm until he was sold after Daly's death in 1900.

The Bitter Root Stock Farm in its heyday encompassed some twenty-two thousand acres and was comprised of several different ranches with multiple horse barns, separate breeding and training facilities for harness horses and racers, four racetracks, a broodmare barn, a veterinary hospital, and substantial and well-appointed homes for the veterinarian, superintendent, jockeys, and breeding foreman. After Daly's death in 1900, the horses were sold at Madison Square Garden. Much of the land and many of the buildings were also eventually sold. Tammany's Castle was used as a stable for many years but was remodeled into a single-family home in 2008.

Tammany, Marcus Daly's favorite horse

When Daly put Tammany on the train to the New Jersey Suburban Stakes, he said, "If Tammany wins, I will build him a castle."

CHAPTER 4

Barns of the Great Divide

The Beaverhead Valley

The Beaverhead is a broad valley crisscrossed by braided creeks and meandering sloughs. Located due north of Fort Hall on the Oregon Trail, the Beaverhead Valley was the portal through which the first herds of livestock entered Montana Territory. This early transportation corridor along the Corinne Road (and later the Utah & Northern rail line) put Montana stock growers on the map.

It all began with substantial gold discoveries along Grasshopper Creek in 1862, discoveries that yielded rich deposits of gold and silver and gave rise to the boom-and-bust mining camp of Bannack. The gold strikes touched off a stampede, with hopeful prospectors flowing in from other mining camps in the West. By the following summer, some three thousand to five thousand miners were busily shoveling, panning, and sluicing their way through the Grasshopper diggings, and in 1864 the young town became the first capital of the newly created territory.[1]

The camp that grew at Bannack was a classic jumble of crude buildings, with dugout shelters, log cabins, and false fronts jockeying for a place along the main road, and barns and outbuildings scattered to the rear. False fronts gave dignity and a sense of style to simple log buildings and, as architectural historian Kingston Heath observed, they were frontier towns' way of "striving for permanence."[2] The few barns remaining in Bannack today, probably built for horses, are among the oldest in Montana. In contrast to storefronts and cabins that incorporated more-finished log work, the surviving barn designs provide relatively crude examples of English-style barns and square-notched joinery. Like most frontier architecture, they were modest in scale and built with local materials.

In 1863, gold strikes in Alder Gulch, east of Bannack and the Beaverhead, set the miners in motion again. More than three thousand took up residence there, with perhaps an equal number scattered in the surrounding gulches and hills. The town of Virginia City took on a unique shape, as described by Samuel Bowles in 1869: "The town lies like a huge serpent, a crooked, irregular strip of log houses winding for nine miles down Alder Gulch."[3]

Nearby valleys attracted farmers and ranchers, and the well-watered grasslands of the Beaverhead Valley became

(opposite) Inside the Barber/Ferris Barn, Madison County

In this circa 1900 view of Bannack, a number of barns sit behind the main buildings. Most were hastily built single-crib structures with a gable roof and log walls. Only two of the early barns still stand today.

important for production of wheat, barley, produce, cattle, sheep, and horses. Livelihoods centered around supplying miners with food and livestock. Many early producers from the Deer Lodge Valley, including Conrad Kohrs, Hank Crawford, and the Stuart brothers, sold beef and other provisions directly from wagons on the street.[4] The trade gave rise to some of Montana's most famous ranches, sprawling, open-range operations, including the Poindexter and Orr (P&O) and the Bishop ranches, whose early log barns are still in use today (see pages 70 and 72).

This agricultural region was accessed by a number of early transportation routes. From the south, the 1880 Utah & Northern Pacific Railroad followed the path of the Corinne Road to Dillon to become Montana Territory's first rail line. An important link connecting Montana to outside markets, the Dillon railhead served as a gateway to a vast rural hinterland—early on, Beaverhead County alone was comprised of 8 million acres. By 1910, the Dillon railhead served an estimated eleven thousand people throughout southwestern Montana and the Lemhi region of Idaho. The rails and stockyards were a bustling scene, and planted conspicuously at the head of Railroad Avenue was the many-gabled P&O barn, a clear symbol of the centrality of stock raising in the young town.[5]

Beaverhead County's high-valley environment was well suited to pasturing livestock, and ranchers in Beaverhead country have consistently raised some of the state's best horses. Early on, most were good-quality common stock, but increasingly, purebred stallions were imported to upgrade the bloodlines. The *Directory for Dillon City and Beaverhead County* of 1916–1917 boasted that "the best grade of horses in the world are to be found here."[6]

The bloodlines reached back to the 1870s, with some of the early breeders having connections to Kentucky and the bluegrass equine industry. Large horse farms were established at the headwaters of the Ruby and Madison rivers, and some of Montana's finest barns were constructed in those valleys. James Mauldin, the first large-scale breeder of blooded workhorses in the territory, bred Normans with his prize stallion, Louis Phillippe, at the Willowburn Ranch east of Dillon, which is known today as the Diamond O Ranch (see page 76). In the Ruby Valley, W. H. Raymond's Commodore Belmont sired a string of trotters at the Belmont Park Ranch, and by the mid-1880s Raymond was one of the largest breeders in the country (see page 100). Deer Lodge banker and mine owner Charles X. Larrabee started a fine Morgan breeding program in 1890 at the Brook-Nook Ranch near Dillon, raising paired teams for carriage and buggy. Best known of all, the Doncaster Round Barn (see page 98) at Noah Armstrong's racing stable near Twin Bridges was home to Spokane and other Thoroughbreds, all trained on Armstrong's private racetrack.[7] The barns on these ranches stand out architecturally in the history of Montana stock raising. In contrast to many log barns of the day, their elegant lines, stick-built framing, and more formal design reflect the investment and aspirations of the early horse breeders of this region.

The Beaverhead was also a center for sheep ranching well into the 1970s. The 1,500 head of sheep that John Bishop trailed in from The Dalles, Oregon, to the Bishop Ranch in 1869, along with 2,700 head trailed in the following year from California to the P&O, lit a fuse under the industry, and it virtually exploded. Sheep numbers grew from roughly 2,000 head in 1870, mainly in the Beaverhead, to almost 280,000 in the southwest and south-central portions of Montana just ten years later. During the early 1900s, 200,000 sheep, 70,000 head of cattle, and 16,000 horses grazed in the Beaverhead, making the county number one in the state in cattle and sheep production and third in horses.[8]

The sheer numbers and diversity of the region's livestock economy resulted in an impressive number and variety of barns being constructed over the years. The Beaverhead features every kind of barn, from single-crib stallion barns, stout workhorse barns, and gambrel-roofed dairy barns to sprawling sheep barns in which thousands of rams and ewes were dipped and shorn each year.

Montana's dry, cold weather and long winters challenged early agriculturists, who learned that growing and storing hay and grain was critical to the success of their ranches. Hay barns and large granaries were requisite technological advances. The early blossoming of stock growing in the Beaverhead led to much innovation and adaptive design that was essential to transplanting agriculture into the northern Rockies. One innovation, widely admired for its practicality and ingenuity, is the beaverslide haystacker.

Spoilage on the outside of a haystack in the pasture encouraged ranchers to stack hay high and keep it dry. Stacking hay became a feat of technology that included derricks with swinging arms and hay ramps and slides that were horse-powered to get the hay into a high stack. Such sliders were in use in the Big Hole region of the Beaverhead by the early 1890s. In 1908, two enterprising designers in the valley came

C-D Ranch hands stack hay with a beaverslide, technology developed in the Big Hole region of the Beaverhead Valley by the early 1890s.

up with a new variant of the hay slide that stacked higher and did not require a long plunger. Use of this more mobile and versatile stacker quickly spread, and the beaverslide, named for the valley of its birth, was soon seen across the Northwest, from Montana and Idaho to Wyoming, Colorado, northern California, British Columbia, and western Nebraska.[9]

The Deer Lodge and Helena Valleys

The "Deer House Plains" is an ample valley and one of the earliest settings for Montana agriculture. Named for the warm-springs mound near the head of the valley, it was a gathering ground for Native peoples. Fur trader W. A. Ferris traversed the Deer Lodge Valley in 1831 with the Pend d'Oreille and Salish and their three thousand horses. Two decades later, a French trader, Jacques Faquaere, drove fifty head of cattle into the upper valley from California, and in the late 1850s livestock growers from the Beaverhead withdrew to the Deer Lodge Valley as trouble between the U.S. Army and Mormon settlers threatened to spill over into neighboring operations.[10]

Métis trader John Francis "Johnny" Grant was the first permanent settler in the valley, staking his future on the emigrant trade and a settlement he named "Grantsville" at the confluence of the Little Blackfoot and the Clark's Fork rivers. When trader James Harkness passed through in 1862, he noted in his diary, "We crossed the Deer Lodge River, a wide and fine stream at this point. . . . I saw several hundred cows and calves belonging to [John] Grant, the finest I have seen in America."[11]

The gold rushes that followed the 1858 gold strike drew population to the valley. By 1863, Grant's ranch spilled over with four thousand head of cattle and more than two thousand horses.[12] Deciding the country was becoming a bit too crowded, Grant sold his ranch to Conrad Kohrs and returned to his home country of Canada.

Kohrs and his half-brother John Bielenberg expanded the ranch to rank among the largest in the United States. The Kohrs Ranch was constantly on the frontier of Montana's livestock operations as stockmen pushed eastward to open ranges in the Sun River Valley, central Montana, and the eastern plains. The brothers strengthened their herds with imported purebred Durham Shorthorn cattle and introduced the first purebred Herefords to Montana's cattle industry. Kohrs also bred livestock for ranch work, and the stallion and oxen barns dating to early years on the ranch are good examples of no-frills buildings suited to their function (see page 140).

The Deer Lodge Valley encompasses some of Montana's earliest developed rangeland, and the barns built here include important examples of log-construction methods and early

The first barn at the Grant-Kohrs Ranch in the Deer Lodge Valley was a "fine, large barn, with a quantity of hay," built in 1865 by Johnny Grant. It was destroyed by a fire within months of its construction, and shortly thereafter Grant sold his place to Conrad Kohrs and moved on. In the decade that followed, Kohrs added the stallion barns (see page 141), draft horse barn (bottom), and oxen barn (top) pictured here. All were photographed in 1974.

agricultural architecture. Impressive log barns that were part of this important early barn building tradition (see page 91) remain today near Drummond and Jens.

In 1864, gold discovered across the mountains to the east sparked a rush to Last Chance Gulch, and the camp, soon known as Helena, grew rapidly. As word spread of the rich ores and extensive minerals in the surrounding foothills, thousands of prospective miners were drawn in. Once here, many turned to other trades to make their livings—carpentry, farming, and commercial ventures. During the 1860s, the Helena Valley, conducive to cool-weather crops, was cultivated for farming and livestock. By 1870, the census noted over a hundred farms in Lewis and Clark County, which led the territory in production of peas, beans, potatoes, butter, and milk.

The mining camps drew prospectors from all over this continent and from foreign lands. People of no fewer than thirteen nationalities—from Europe to China and Japan to the West Indies and the Sandwich Islands—were noted in early population records. The communities that grew up around the mining camps drew their share of ethnic settlers, and as they did, ethnic farmsteads and barns in the region came to reflect the ideas and farming customs traditional in the home country of the settlers. Croatian and Italian stone barns in Helena and English-, Irish-, and German-type barns in the Helmville Valley are among the early ethnic barns of the area (see Jezick Barn, page 119).

Helena drew self-made late-nineteenth-century capitalists who expressed their growing prosperity in high-style architecture in the city, while the Helena Valley developed livestock and crop production. By 1890, the valley held over three hundred farms and ranches. Most outstanding were cattle ranches owned by brothers William C. and Harry W. Child, who built their fortunes through mining, ranching, and the Yellowstone Transportation Company. Their Green Meadow and White Face ranches (known today as the Kleffner Ranch, see

Designed by Robert C. Reamer, the buildings on Helena's Green Meadow Ranch were built in the Bavarian Swiss style with heavy-timber roof supports, half-timbered walls, and rolled shingle roofs. The barn was 450 feet long, had three drive-through aisles, and in addition to horses and tack, housed up to sixty white-faced Herefords at any given time.

page 128) featured the valley's most impressive and massive barns.

Through the twentieth century, farming continued as an important part of the greater Helena economy. Dairy, cattle, and sheep barns remain today and reflect a direct line of agriculture that ran from farms in the valley to the markets and people in town. The Sieben Ranch, one of Montana's largest sheep ranches, still operates, and its century-old sheep sheds and barns are important examples of buildings that have not lost their purpose (see page 120).

The Missouri Headwaters

The Missouri River connected settlements in the region where its headwaters come together and then flow on as one. Downriver from Helena, the Great Falls of the Missouri blocked river navigation, leading traders to establish riverside posts in the 1840s and later, steamboat landings below the falls at Forts Benton and Carroll. Upriver, where the Jefferson, Madison, and Gallatin rivers flow together to form the Missouri, John Colter and George Drouillard of the Lewis and Clark Expedition established one of Montana's first trading post in 1810. The post was short-lived, but hopeful settlers returned in 1862, constructing a cluster of log buildings at the headwaters that they ambitiously named Gallatin City.

The valley's biggest town, Bozeman, was named for John Bozeman, who blazed a route up from the Oregon Trail in Wyoming to the Gallatin Valley and beyond. Those traveling through could stop over near Fort Ellis and then continue on to the goldfields, turning left for Virginia City or taking a hard right to follow the Missouri downriver toward Helena. The Bozeman Trail crested the mountains at Bozeman Pass and then dropped to the valley floor, where, in the mid-1860s, there was just a handful of log cabins. From the pass, newly arriving immigrants were able to take in the verdant Gallatin Valley, destined to become the breadbasket of Montana.

Yearly rainfall varies dramatically across this valley, ranging from a marginal eight inches on the east near the Horseshoe Hills to sixteen inches on the west end at Bozeman. Weather conditions can be harsh at this mile-high altitude, as William Alderson noted: "Frequent showers of snow and such hard freezing at nights occur that it is really discouraging to a person expecting to make money farming here." Still, it was John Bozeman himself who noted in the 1860s that the valley was "fast being settled up with farmers, many of whom came to Montana as a better class of miners." Across the valley, a familiar story of farmers serving miners unfolded as the first grains, potatoes, and vegetables were harvested and irrigation ditches were dug. In 1865—as the town of Bozeman celebrated its first birthday—fifteen hundred acres of cultivated land grew twenty thousand golden bushels of wheat.[13]

Nearby, other drainages were also farmed to provision the mining camps. The Willow Creek area, west toward Virginia City along the Bozeman Trail, was a protected valley that was settled early and was, in fact, the point of origin from which all federal land surveys in Montana Territory were measured. The farms here were among the earliest in this region, and many included modest, first-generation log barns. While many of these were replaced as the valley grew, they can still be spotted today, now storing machinery and sometimes hay.

The Gallatin was a fertile plain for growing grains, and Belgrade turned into one of the largest wheat-shipping hubs in the state. Each year, tons of golden Gallatin grain made their way to market, giving rise to P. D. McAdow's Mill, which was famed for producing a thousand sacks of flour a week, and its competitors, the Penwell brothers' Union Flour Mill at Springhill and Nelson Story's Story Mill up Bridger Canyon. A miner who had struck pay dirt in Alder Gulch, Nelson Story also pursued stock raising, trading gold dust for cash to purchase Texas longhorns. From the Guadalupe River to Montana via the old Texas Trail, Story's longhorns were the first to reach Montana territory in 1866–1867.[14]

Cultivation of the Gallatin expanded dramatically after the arrival of the Northern Pacific Railroad—and was encouraged by Henry Altenbrand's scheme of the early 1890s to construct a large-scale irrigation system in the Gallatin. Altenbrand, a brewer from Brooklyn, New York, planned to develop industrial malting facilities at Manhattan. He and his investors

The Gallatin Valley was one of the largest wheat-shipping hubs in the state. Some of that wealth was stored in this granary, pictured here circa 1910.

In 1919, Prohibition dried up the grain market, and the Dutch farmers, who had come to the Gallatin Valley in the 1890s, diversified and produced prodigious amounts of potatoes, hogs, and dairy products. Here, George Diest poses in a potato field.

obtained title to thirteen thousand acres, dug a one-hundred-mile canal network, and recruited Dutch farmers to settle, plant barley, and malt grain. In the words of one early settler: "I never saw such fields of grain and other produce. . . . In every direction could be seen fields of golden and heavy grain still in the shock."[15] For two decades, the New Hollanders prospered, growing oats, wheat, and barley and thriving in the pre–World War I market. Prospects changed—literally dried up—with passage of the Eighteenth Amendment and the National Prohibition Act of 1919. Dutch farmers diversified, and the Gallatin's prodigious production of potatoes, hogs, and dairy products was centered in the New Holland settlement through the later twentieth century. Barns there reflected an orderly and up-to-date approach to farming; wood-frame buildings with one-and-a-half- to two-story gambrel and monitor roofs afforded plenty of storage space.[16]

Irrigation also allowed dairy operations to expand in the east Gallatin, supplying the growing city of Bozeman with locally produced cream, butter, and cheeses. Throughout the Gallatin Valley, many large, second-generation dairies bred dairy cows that required shelter, feed, and twice-a-day milking. By this time, manufacturing and railroad expansion offered a multitude of building supplies. Spacious barns, large haylofts, silos, and milking parlors are the hallmarks of twentieth-century dairy farms.

Montana State College of Agriculture and Mechanic Arts was established in Bozeman in 1893, a complement to all of the agricultural activity going on around the headwaters region. The college's first building, Taylor Hall, housed the state's agricultural experiment program. Befitting its mission, Taylor Hall was built in the shape of a barn and was a center of progressive agriculture, promoting new ideas and research on crops and livestock production. For example, in 1910 the college planted the first seed peas in the valley, and by 1913 seventeen thousand acres and four seed pea companies were flourishing. Drawings for two seed pea warehouses by prominent architect Fred Willson are among the drawing collections housed today at Montana State University.

Over the years, the agricultural college grew, and the impressive array of barns designed by Professor William Plew and others demonstrated modern ideals of barn construction and progressive approaches to farming and animal husbandry. The university's agricultural experiment program published scores of circulars on farming techniques and farm building design. Instructions were given for building everything from ice houses to poultry houses to dairy barns. Agricultural engineer H. E. Murdock expressed the optimistic philosophy in 1926 that

> With the rapid development of agricultural lands and the division of the large stock ranches of Montana into smaller farms there will come large investments in buildings. . . . When livestock is kept, shelter of some kind must be provided, and the construction of barns represents a heavy permanent outlay.
>
> Building a house or barn is an event that usually occurs once or twice in the lifetime of a farmer and it should be given careful study beforehand. . . . The buildings on the farm should express simplicity, freedom, and rest, and their architecture should be along these lines.[17]

Leading Montana architects including William R. Plew and Fred Willson designed impressive early-twentieth-century barns for the Montana State agricultural program in Bozeman. In later years, their sites were taken by new university buildings and the Museum of the Rockies.

Poindexter and Orr Barn

BEAVERHEAD COUNTY, CIRCA 1870s

This back view of the Poindexter and Orr barn clearly shows the original pitch of the wide gable roof, as depicted in the historic 1885 lithograph (far right, bottom). The monitor roof was added in the early 1900s to increase storage space in the loft.

The Matador Cattle Company still uses the original Poindexter and Orr brand—the first brand to be registered in Montana.

Built in the 1870s or early 1880s as the main horse barn on the Philip H. Poindexter and William C. Orr Ranch, the P&O Barn is associated with one of the earliest and most important stock-raising operations in Montana. The Poindexter and Orr Cattle Company was the first outfit to trail in cattle to the Bannack area in 1864, officially establishing the cattle industry in the state. In just a few short years, the P&O Ranch became one of the largest stock-raising operations in the state and remained so for more than fifty years. Since then, the sturdy log barn has been continuously used to house horses and hay.

A closer look at the exterior of the barn reveals a log building of a size and type of construction seen only on some of the most successful ranches in Montana. Originally, the barn had a wide, low-pitched gable roof (as seen in the 1885 lithograph at bottom right), and the logs would have been exposed, not covered with siding, and likely left unpainted. When viewed from the rear, the barn reveals the original pitch of the roof and the hand-hewn square logs with fine half-dovetail notched corners. The barn is about seventy feet long and sixty feet wide, consisting of four consecutive log sections, huge compared with the simple one-room or single-crib horse barns of many early settlers. Sometime in the early 1900s, a central monitor roof was added to expand hayloft storage, and lapped wood siding and red paint were added to give a finished look to the façade.

Little is known about the carpenters who built the barn, but once inside, even the casual observer cannot help but soak up history in the walls, the floors, and even the doors. The "Fool's Register," a cacophony of cowboy graffiti in the form of carved and written initials, names, dates, drawings, and even weights and measures of horses on the tack room doors and walls, is a fascinating record of the many ranch hands who walked across the barn's massive split-log floors on their way to saddle a horse. The oldest graffiti dates to the 1880s.

(above) The "Fools Register," inside the P&O horse barn

(left) Leeson's 1885 *History of Montana* clearly illustrates the Poindexter and Orr barn before the addition of the monitor roof.

Bishop Ranch Barn

BEAVERHEAD COUNTY, CIRCA 1870

The Bishop Ranch represents the beginning of the sheep-ranching industry in Montana. John Bishop, one of the earliest settlers in the Beaverhead Valley, drove in the first band of fifteen hundred sheep from The Dalles, Oregon, in 1869. The eight-hundred-mile trip took eighty days and much perseverance, but Bishop was able to sell the wool to Charles A. Broadwater in the spring for nineteen cents a pound, a handsome payoff. Nearby ranchers Poindexter and Orr drove in another band of sheep in 1871, and the sheep industry in the valley expanded rapidly. Beaverhead County remained one of the largest sheep- and wool-producing counties in the state for many years, and during the 1920s and 1930s, it had one of the largest wool-shipping warehouses west of the Mississippi.

Though the roof on the Bishop Ranch Barn was enlarged from a smaller gable roof to a gambrel roof, the barn depicted in the 1885 lithograph (far right, bottom) is the same one that still stands on the ranch today. This type of mountain horse barn pictured in the lithograph was at one time a common barn type on many of Montana's early ranches. The central gable-roofed section was typically one room with a central alley and stalls lining the walls. There was a hayloft above, and the side wings, which were sometimes added later, were used mainly for storage. These early barns were almost always constructed of squared or rounded logs. Knowing the simple characteristics of these rough-hewn barns, the casual traveler on Montana's rural roads can easily identify ranches and farms associated with early stock-raising activities.

John Bishop brought in the first large band of sheep to Montana Territory in 1869 and started the sheep operation that became the two-thousand-acre Silver Lake Ranch depicted in this 1885 lithograph.

John Bishop brought in the first large band of sheep to Montana Territory in 1869 and started the sheep operation that became the two-thousand-acre Silver Lake Ranch depicted in this 1885 lithograph.

Selway Barn

BEAVERHEAD COUNTY, 1906

The Selway Barn was built for E. O. Selway in 1906 by master barn builder John Kent, a well-known builder of many barns in the Beaverhead and Madison valleys, including those at the DL Tash Grasshopper Ranch at Polaris and the Diamond O Ranch as well as some of the Poindexter and Orr (P&O) Ranch buildings (pages 90, 76, 70). Kent's barns are easy to spot as most are early-1900s frame structures and sport distinctive "flying diamond" windows set into their red walls.

E. O. Selway at one time ran nine thousand head of sheep in the Beaverhead Valley. With such a large operation, Selway and his crew would travel several times a year to Dillon with teams and wagons to haul the wool and pick up supplies. The men would make it to the P&O Ranch on the first night, then to town the next day, with another night back at the P&O before reaching the home ranch on their return.

The handsome three-story barn housed sheep on the first floor, and during Selway's era, oats were stored in bins and wool on the second floor. An elevator in the center of the barn was used to hoist wool bags into the second story. After Selway sold the ranch, one owner used the second floor for lambing, driving sheep up to the second level on a ramp and later bringing them down again on the elevator platform.

Diamond O Barn

BEAVERHEAD COUNTY, CIRCA 1870

The history of the Diamond O Ranch near Point of Rocks spans back to the early days of stock raising in Beaverhead County. James Mauldin came to Montana in 1865 and made his living selling shoes to miners in Virginia City and other mining camps. He got out of the shoe business in 1868, trading his footwear inventory for cattle and trailing them from Utah to the Beaverhead Valley. One year later, he brought five hundred Norman horses from California and began the Willow Burn Stock Farm with the Diamond O brand. He also brought to the valley the first Percheron horses west of the Missouri River. Over the next twenty-eight years, Mauldin contributed greatly to the breeding of cattle and workhorses in southwest Montana. He retired from ranching in 1898 and sold out to Ira Sanders and A. D. LaDue, prosperous cattle barons in the region at the time.

The barn was likely built for Mauldin circa 1870 and later remodeled in the early 1900s by Sanders and LaDue. An 1885 lithograph depicting Willow Burn Stock Farm shows that the barn originally was sided with wide vertical boards and had a single roof-top ventilator centered on each wing. In addition, the right-hand wing of the barn once held four bays on the lower level for wagon and carriage parking. Current pictures of the barn clearly show a "ghost" line where the header spanned the parking area. The only original features remaining on the barn are a row of small square windows on the east side and the hayloft door above the main double doors.

FRIENDS

Staudaher Barn

BEAVERHEAD COUNTY, CIRCA 1900

The Staudaher Barn is an unusually large and unaltered example of a mountain horse barn. The exterior log work is impressive and clearly the work of experienced carpenters. Each log has been carefully chosen and end-hewn for uniform size, and there is no chinking between logs, owing to the tight-fitting square-notched corner joints. The cathedral-like interior of the barn is equally impressive with its heavy-timber framing. It is fairly unusual because it has no hayloft floor above the stalls. Instead, there is a wide central alley, open to the roof for hay storage, and two side alleys lined with horse stalls. Hay could be stacked by hand to a certain height and then stacked higher using horsepower, pulleys, and grapples through the rear hayloft door.

General Land Office maps printed in 1902 list George J. Staudaher as owner of this land. Born in 1865, Staudaher was one of the first babies to be delivered in Virginia City, and he grew up ranching on his parents' Pearl Spring Ranch north of Dillon. He and his wife, Rose, later owned a considerable amount of land in the Beaverhead and Centennial valleys in the early 1900s, and this massive barn reflects their prosperity.

The barn was likely built around 1900 with help from Austrian craftsmen living in the area. Staudaher was a part of a large and extended network of Austrian immigrants to the Beaverhead Valley. His parents, George L. and Catherine, are known to have brought many settlers of Austrian descent to the valley in the late 1860s.

It is rare to find such a large barn without a hayloft floor.

Cook Sheep Company Barns

BEAVERHEAD COUNTY, 1900–1915

The Cook Sheep Company Ranch was one of the largest Beaverhead County sheep ranches of the early 1900s. J. E. Morse bought the Steve Cook holdings on Sage Creek near Dell in 1913 and expanded them by purchasing other Sage Creek ranches, leasing land from the State of Montana, and using U.S. Forest Service permits. By 1919, Morse was running herds of sheep totaling sixteen thousand head. In many years, the Cook Sheep Company won the highest price for fleeces at market. In July 1927, Morse sold 150,000 pounds of wool at 34¼ cents per pound. At the time it was one of the largest wool deals in Montana history.

J. E. Morse was an experienced and trusted businessman in Dillon. He was one of the original founders of Dillon in 1880 and the owner for many years of the Dillon Implement Company. When he set up his ranching operation on Sage Creek, Morse laid it out Main Street style, complete with horse barns and a blacksmith shop, harness shop, implement shed, garage, bunkhouse, and commissary strung out along both sides of a central road. In this vast, sparsely populated area, the ranch operated much like a small town, with the commissary acting as a general store and restaurant to serve the ranch hands and local homesteaders, who otherwise would have to travel rough roads to Dell or Lima for food and supplies.

J. E. Morse died suddenly in 1928, and his son Howard took over the operation. In the 1950s, the Matador Cattle Company bought the ranch and has meticulously maintained the buildings ever since.

The Cook Sheep Company was one of the largest sheep operations in early-twentieth-century Beaverhead County. It was so large that the headquarters ranch was set up like a small town, with a blacksmith shop, a bunkhouse, a commissary, a manager's house, and matching gambrel-roof barns laid out Main Street style.

Forrester Granary, upper ranch

Forrester Granary, lower ranch

Towery Granary

Blacktail Granaries

BEAVERHEAD COUNTY, CIRCA 1900

This collection of nearly identical granaries along the Blacktail Road near Dillon were likely constructed in the early 1900s. This most basic type of outbuilding is referred to as a crib-type granary. Constructed of two-by-fours laid flat and spiked together, this granary could be put up quickly by almost any builder with low-quality or short-length lumber. The stacked lumber provided a higher measure of wall strength than a balloon-frame building, suitable for storing grain in excess of ten thousand bushels. Because of the sturdiness, this construction method was also used on the walls of grain elevators. As an added advantage, these tight walls also helped to keep pests and moisture away. Although these buildings have not been used for grain storage for many years, they remain an important fixture in the farming landscape of the Beaverhead Valley.

Barns on the Marchesseault Ranch represent two eras of agricultural growth in Montana. The circa 1870 log barn (above) is a pristine example of a single-crib log barn built of local materials, while the 1950s concrete block barn (opposite, top) represents the materials and technology embraced by Montana ranchers following World War II.

Marchesseault Barns

BEAVERHEAD COUNTY, CIRCA 1870 AND 1950

In stark contrast to the larger and later barns in the Beaverhead Valley, the circa-1870 log barn on the Marchesseault Ranch near Dillon is a perfect example of a single-crib, or one-room, barn. This type of barn was most common on early ranches in Montana because it was the simplest to construct with abundant local timber. This particular barn is in strikingly good condition and is rare since it still has a wood-slat roof and original doors and window openings and has not been added on to or altered on the inside. Typically ten feet wide and eight feet deep, with a side gable roof and central entrance, small barns like this one are becoming harder and harder to find in Montana.

The origins of this barn are unknown; early land maps show a structure on the property as early as 1872, and General Land Office records reveal that the land was patented by Leobold Born in 1877. The Marchesseault family bought the property in the 1960s and began expanding the ranch and improving the buildings. Today, the old log barn stands in the shadow of two post–World War II concrete-block barns built by the prior owners Marcus and Ann Staudohar—barns that have become historic in their own right.

Owner Jules Marchesseault is proud to carry on the family ranch.

Dewey Twin Barns

BEAVERHEAD COUNTY, CIRCA 1877

These two nearly identical buildings are among the first substantial buildings to be constructed in the small mining town of Dewey. Built by Isaac Dodson in about 1877, the building on the left in the center photograph was a boardinghouse and the other a livery barn for the stage stop.

Looking closely at the historic view of Dewey below, one can see that these buildings have changed only slightly. The boardinghouse had a central door with two double-hung windows above, and the open space between the buildings was

The twin barns (center) stood on Dewey's main street.

Detail of the dovetail notching on the boardinghouse

covered by a shed roof. The barn on the right appears to be unaltered from its original design. It is also interesting to note that although both structures are skillfully constructed, the hand-hewn logs and dovetail notching on the boardinghouse are much more elegant (and airtight) than the irregular and somewhat rough log work on the barn. At a later date, a crude log addition was attached to the rear of the boardinghouse.

Dodson was a millwright in England who came to Montana in the 1860s with the general rush of miners into the territory. In 1875, he and his family were the first to make a home in Dewey. At the same time, Dodson built and began operating an arrastra mill nearby. The livery barn is an excellent example of a barn that would have been found in nearly every burgeoning Montana town in the late 1800s.

TASH
RANCH

Tash Barn

BEAVERHEAD COUNTY, CIRCA 1868

The Tash Barn just outside Dillon is one of Montana's early ethnic barns. Constructed in the mid- to late 1860s, the central building is constructed of hand-hewn square logs in the style called *piece-sur-piece* construction.

The technique uses squared horizontal logs, beveled at the ends, and stacked or slid down between vertical channel posts. The channel logs in the Tash Barn are found midway between the north and south walls and as the doorjambs on the east and west walls, thus extending the wall length. The vertical channels rest on the sill log, and the horizontal logs are held in place with square, hand-forged iron spikes. The sill logs are not continuous but are composed of two logs lapped at the wall midpoint and held together with wooden pegs. At the corners, the logs are notched using a single dovetail joint. This ingenious method of using the channel post as the center post allowed for a larger barn using shorter lengths of log than would be necessary with corner-post construction.

French-Canadian residents Amede Bessette and/or Xavier Renois are credited with the construction of this barn and many of the ranch's other early buildings. Bessette and Renois were also responsible for the construction of several buildings in nearby Bannack, most notably the Mead Hotel.

The dinner bell at the Tash Ranch

Hand-hewn square logs form the walls.

The barn in this 1885 lithograph is the same one that stands today. Amede Bessette and/or Xavier Renois likely built it in the mid-1860s for Isaac Van Camp. The side sheds were added in the early 1900s.

DL Tash Grasshopper Ranch Barn

BEAVERHEAD COUNTY, 1914

The DL Tash Grasshopper Ranch Barn is an outstanding example of a barn built by John Kent of Dillon. Embellishing on an ordinary barn design format—the gambrel roof, center doors, central aisle, and side windows—Kent distinguished this barn with a wide rake board and flared eaves to define the roofline, a tall hexagonal ventilator with a pagoda roof, and his signature "flying diamond" windows.

In 1914, William Tash and partner John Lenkersdorfer had Kent build the barn. Two years later, the main house was built. In 1918, Tash bought out Lenkersdorfer's interest, which included four thousand acres of deeded land, two thousand acres of leased land, and a thousand head of cattle and horses. Tash later became known for raising the finest Hereford cattle in the state. The DL Tash name was devised by reversing the initials for Lenkersdorfer (LD). When he sold out, the name stayed with the ranch.

The "flying diamond" window is the signature element of the barns built by John Kent.

Mule Ranch Barn

DEER LODGE COUNTY, CIRCA 1885

The agricultural buildings still standing in the Mount Haggin Wildlife Management Area were originally part of a large sheep ranch owned by the Anaconda Copper Mining Company. Apparently, A. W. McCune, an early settler in the area in the 1880s, grazed mules used in his logging operation on a 160-acre parcel in the Mount Haggin area. It became known as the "Mule Ranch," and the name stuck even after the Anaconda Company's Deer Lodge Valley Farms Company grazed cattle and horses there. Even later, in the 1920s, when Dr. Henry C. Gardiner, famous for his Hampshire sheep, leased the land from the Anaconda Company, it was still known as the Mule Ranch.

This barn is significant for its association with the Anaconda Company and with Henry Gardiner's successful sheep-raising operation. It is also an important example of the fine craftsmanship found in the area, and the log joinery on its large single-log crib indicates a builder skilled in the use of an axe and adze. The uniform end-hewn logs and double-lapped corners took experience and time to achieve. The result was a very tight-fitting log wall and a warm and hospitable dwelling for work animals.

Wold Barn

SILVER BOW COUNTY, 1907–1909

Ida Wold left no record of who built the barn, but the high-quality craftsmanship of the builder is clear in the hand-hewn square logs and square-notched corners.

(above) A hand-forged hitching loop still hangs on the outside of the barn.

(right) Steep stairs lead to the hayloft.

Ida Longley Wold—wife, mother, and Melrose business owner—was an independent woman and the driving force behind the construction of this barn.

Ida Longley married Oscar Wold in the 1890s, and the couple moved to Melrose, where Oscar was a teamster. By 1900, they had three children. In 1901, Ida purchased ten city lots, and in 1904 she opened a boardinghouse there. An article in the December 5, 1905, *Anaconda Standard*, titled "Ida Wold Her Own Boss," stated that "Judge Bourquin yesterday granted Ida M. Wold of Melrose sole tradership papers, and hereafter the woman will conduct her business in her own name. She conducts a boarding house in Melrose, and since launching the business 15 months ago has acquired five horses and two cows from the proceeds of a $600 investment. She has three children to support."

The news item suggests that Ida Wold was the primary wage earner for her family, and while there is no mention of a barn, she would have needed one to shelter her newly acquired livestock. Early photos do not show the barn until after 1906, suggesting that Ida either hired a builder or perhaps her husband built the barn between 1907 and 1909, the year that Ida passed away at age forty-eight. A later article in the *Standard* related that "During her lifetime Mrs. Ida Wold of Melrose was the owner of a lot of horses and cattle, which are now under the control of her husband, Oscar Wold." The brief report goes on to tell of a creditor who tried to settle things with Mr. Wold—unsuccessfully, since "Wold was intoxicated most of the time and was not looking after the property."

While many barns in this book tell of the successful progression of men in ranching and farming, the story of the Wold Barn offers a glimpse into the experiences of an independent woman who resourcefully navigated the worlds of business and farming to support her family.

Madison County Fairgrounds Barns

MADISON COUNTY, 1936–1937

(left to right) Horse stables; exterior of the pavilion; one of three WPA-era barns at the fairgrounds; pavilion interior

The collection of historic buildings at the Madison County fairgrounds in Twin Bridges is an outstanding example of Works Progress Administration (WPA) design and construction using entirely local labor and materials. In addition to the impressive octagonal pavilion, grandstands, caretaker's house, and bandstand, there are three distinctive log barns that were part of the fairgrounds WPA-era expansion begun in 1936.

The fifty acres that compose the Madison County fairgrounds were once privately owned by the Lott and Siedensticker families and were developed as a community park beginning in 1887. The fair thrived through the 1890s and early 1900s as a gathering place for local farmers and ranchers to show their produce and livestock while promoting agricultural development in the region.

Construction on seven new buildings began in 1936 and was largely completed by 1937. All of the barns were designed in the rustic style by WPA engineer C. D. Paxton with construction directed by Tosten Stenberg, who was well known for his work on large log buildings in Yellowstone National Park. Today, the rustic log barns and community buildings at the Madison County fairgrounds continue to symbolize WPA architecture of the 1930s and its importance in reviving rural economies.

Swanstrum Barn

MADISON COUNTY, EARLY 1870s

Though slightly hidden on the landscape today, there is little doubt that the Swanstrum Barn was a well-known landmark on the wagon road from Twin Bridges to Dillon in the late nineteenth century. The barn, constructed in the early 1870s, must have drawn attention as one of the most picturesque outbuildings in rough and rugged Montana Territory. While many barns of the 1860s and 1870s had a simple gable roof, rough round logs, and one small room to house horses, tack, grain, hay, and equipment, this barn, with its steeply pitched cross-gable roof, hexagonal cupola, and uniform siding and trim, stands apart from all other barns in the area.

The interior was more spacious and commodious than most barns of the day and certainly more comfortable than the original homestead cabin nearby. Like the Tash Barn (page 88) near Dillon, this barn was wide enough to have a central alley, which allowed for access to stalls on one side and various storerooms on the other. Upstairs, the already tall hayloft was enhanced by the steep, pointed side gable. In

examining the exposed log work at the rear of the building, one can see that short lengths of logs were cleverly woven together in order to make a much wider barn—a construction detail that required some engineering know-how.

The distinctive form, masterful log joinery, and detailed finish work on this barn are the work of Swedish immigrant Carl Swanstrum. Swanstrum trained as a carpenter in Sweden and came to the United States in 1864. By 1865, he was at work building railroad bridges throughout Iowa. In 1871, he moved to Montana and later patented a claim of 160 acres in Madison County.

The steeply pitched roof of this barn suggests the Gothic Revival style, popular in this country between 1840 and 1880, primarily promoted by Andrew Jackson Downing's books for cottage residences and country homes. These publications stressed the suitability of the Gothic Revival form as a rural style, compatible with the natural landscape. In his *Cottage Residences,* first published in 1856, Downing promoted a plan for a carriage house and stable in the "rustic pointed style." Though far more ornate than the Swanstrum Barn, the rendering shows a barn with board-and-batten siding, a pointy cross-gable roof, and a steeple-like cupola. Swanstrum may have been influenced by such design ideals, or perhaps the lines of the barn derive from a building tradition of his native Scandinavia.

Doncaster Round Barn

MADISON COUNTY, 1883

The much-beloved Doncaster Round Barn near Twin Bridges is one of the most famous barns in Montana. It was praised in horse-breeding annals and is considered an outstanding example of the rural architecture of the state. Many great horses were foaled and stabled under this circular roof, most notably Spokane, who claimed an upset win at the 1889 Kentucky Derby—the only Montana horse ever to win that crown.

The barn appears to take its inspiration from the great Kentucky horse farms of the late nineteenth century. Noah Armstrong, a native Kentuckian, made a fortune mining and smelting in southwest Montana, then began to pursue his passion for horse racing. In the early 1880s, he found perfect grassland at the base of the Tobacco Root Mountains, acquired over four thousand acres near Twin Bridges, and named the ranch Doncaster after his black Thoroughbred stallion.

As Charles Armstrong, Noah's son, later wrote: "A good sized fortune has already been expended in erecting the buildings and accommodations essential to the proper housing and handling of the valuable animals which are domesticated on Doncaster Ranch and for their care and maintenance in a manner befitting their lofty equine station. No provisions for their well-being are lacking; the buildings are convenient and substantial; [and] the sanitary condition of the barns and stables has been the subject of due consideration."

The stable floor was one hundred feet in diameter, with offices and sleeping rooms flanking the main door, a central water well, and twelve-foot-square box stalls along the wall. The second floor was used for hay and grain storage, and the third floor ingeniously housed an eleven-thousand-gallon circular water tank. Water was pumped from the well to the storage tank by a windmill on the roof and then gravity-fed

through pipes down to the horse stalls. Indoor plumbing was a novelty in Montana in 1883, and this barn was likely the only one of its time to boast such a convenience.

In addition to the interior conveniences, each horse stall had a large window and a door to a separate two-and-one-half-acre paddock. Nearby, horses trained on a regulation half-mile track. Charles Armstrong also quoted neighboring rancher E. S. Larabie as saying: "I am quite sure that, for elegance of design, convenience in all its appointments and general adaptability . . . there is nothing equal to it on any of the great horse farms of Kentucky."

Although many of the great sprawling and steepled horse barns of Kentucky were not round, agricultural scientists in the mid-1800s began to promote the use of round barns, extolling the virtues of their practicality, efficiency, and cleanliness. Many round barns were used for dairying, but a few housed horses. In 1882, a four-story round barn was constructed in Lexington at the Kentucky Fair Grounds as a floral exhibition hall. A few years later, it was converted for use as a horse barn by the Kentucky Trotting Association. It later became famous as the stable for the Red Mile Trotting Track, where harness racing was fully developed. Apparently, the Doncaster Round Barn was conceived and designed by Armstrong himself; perhaps he was influenced by this prominent round barn in his native state.

Noah Armstrong's Doncaster Round Barn appears to take its design inspiration from some of the great Kentucky horse barns of the nineteenth century. Each horse stall has a window and a door that once led to a separate two-and-one-half-acre paddock.

Handlers show Belmont Park horses in front of their barn.

Belmont Park Barn

MADISON COUNTY, CIRCA 1880

In the early 1880s, William Hillhouse Raymond brought a taste of Kentucky to Madison County at his Belmont Park Ranch on the Sweetwater River. This affluent-looking barn, with its ornate steeple-style cupola and well-appointed interior, is strongly reminiscent of more palatial horse barns found near Lexington, Kentucky, the horse capital of the world. Along with the famous Doncaster Round Barn (see page 98) farther north, the two barns represent the imprint of high-style Kentucky barns upon the legacy of horse ranching in southwest Montana.

Raymond began amassing his fortune in Montana in 1865, freighting grocery goods from Corinne, Utah, to Virginia City. In the early 1870s, he and his brother Winthrop developed a horse ranch near the Sweetwater Bridge. As the owner of one of the first and most successful horse ranches in the state, Raymond was a frequent visitor to Kentucky, and in 1876 he took an extended trip there to purchase his ideal stallion, Commodore Belmont,

Commodore Belmont (center)

and twenty broodmares. In the following years, Raymond imported dozens more horses from Kentucky, including two Shetland ponies, the first Shetlands in Montana.

In the early 1880s, Raymond again expanded and built a training facility, which he called Belmont Park after his favorite horse and the famous New York track. At this ranch, he built a handsome mansion, the showy white horse barn, sheds, corrals, and a one-mile racetrack. Horses such as Carrie Belle, Doncaster, and others well known on Montana tracks trained at Belmont Park. Raymond also farmed, mostly raising feed for his stock. Interestingly, he did not use workhorses for field work but instead harnessed up a span of colts to the plow to increase their endurance on the track.

Belmont Park was reportedly the first racehorse stable in the Northwest, with the largest number of Standardbred trotting horses in the world in the 1880s. The breeding empire was dissolved when Raymond died in 1905. The ranch sold in 1910 to Andrew and Joe Williams, who raised cattle, sheep, hogs, and Belgian draft horses. The barn still stands majestic today, a stately reminder of Raymond's legacy and his career in raising and breeding horses in the Northwest.

Foster/Davis Ranch Barn

MADISON COUNTY, 1877

This stately stone barn dates back to 1877 and the establishment of the small community of Waterloo in the northernmost part of Madison County. Originally built as a woolen mill on land donated by Zenas Foster, the structure was erected by stonemason Pat Carney using granite from the nearby hills and heavy posts and beams. Several other early settlers in the area provided capital, and when the mill was complete, the investors threw a big party, only to have a gust of wind blow the roof off in the middle of the bash. The roof was replaced soon after, but it blew off once again. With no roof and waning support from investors, the mill sat empty.

The springhouse reflects the style of the barn.

After 1877, the barn may have been used for soap making by Zenas Foster's wife, Emily, whose soft-beaver-fat-and-lye soap making was a successful pioneering business. The Fosters sold out in 1893 to wealthy Butte banker A. J. Davis, the nephew and namesake of Butte's first millionaire mining magnate, A. J. Davis.

Davis added a sturdy roof to the barn and retrofitted the interior for use as a horse barn. He spared no expense on the comfort of his horses and ranch hands. The interior stalls stand independent of the stone walls and were framed using heavy timbers connected with fine mortise-and-tenon joinery. Even the hayloft railing is neatly turned and curved, a detail not often fussed over in a barn. Nearby, a matching granite springhouse and generator building were erected along Spring Creek as well as a handsome wood-frame carriage house. The Davis family lived most of the time in Butte, using the Waterloo Ranch as their country place, a weekend and summer playground at which family and friends gathered.

Barber/Ferris Barn

MADISON COUNTY, CIRCA 1865

Current owner Kathleen Wuelfing and her father, Jack Fenton, pose in front of the old sheep barn. Wuelfing grew up on the ranch and recently restored the homestead house and barns. The barns were stabilized with grant funds from the Montana State Historic Preservation Office.

Standing aglow against the early morning sun, the humble Barber/Ferris Barn near Sheridan is a remarkably well-preserved English-style barn dating to the mid-1860s.

The log portion of the barn was most likely built by Canadian stock-raiser John Barber around 1865. The rubble-stone foundation, peeled round-log walls, saddle-notched corners, and wood-slat roof illustrate typical materials and craftsmanship of the 1860s. The slat roof is now a rarity, but it was the roof of necessity in territorial days when hand-hewing shingles and installing them on such a steep pitch would have required countless hours and a skilled carpenter. The wood-frame addition to the rear was built between 1883 and 1937.

If the logs and hand-forged iron hardware on the barn could talk, they might tell the brave story of Jane Ferris, an early owner of this ranch. A widow with two children, she was one of the few pioneer women in Montana Territory to claim land in the early 1870s, a time when many women would have been resistant to the idea of moving to the West, even when accompanied by a husband, relatives, or friends. Ferris, however, did it with only the support of her young children.

Back in 1864, en route to Virginia City to join her husband, Mrs. Ferris got word that he had been killed in a mining accident. She briefly returned to Colorado but made her way to Sheridan in 1866 and took work as a housekeeper for John Barber. In 1872, Barber died, and Ferris was able to claim 160 acres of his ranch under the 1841 Preemption Act. Her application notes a cabin and barn already on the property, which had a value of five hundred dollars.

Ferris appears to be the first woman in Sheridan-area history to use preemption to secure land. Sadly, she died in 1873, taking the story of her journey with her but leaving the land to her son and daughter. Her daughter later returned to the ranch in the late 1870s and continued ranching with her husband until 1883. Although Ferris's tenure in frontier Montana was short and her full story unknown, her experience sheds light on the ways that women came west, claimed land, and made a new life with their families.

Show barn

Horse barn

Jumping Horse Ranch Barn

MADISON COUNTY, 1929

Ranch architecture in the Madison Valley was elevated to a new level in 1929 when Dorothy Hodges, her husband Wetmore, and Swedish carpenter Jack Hill developed the Jumping Horse Ranch. After purchasing several valley ranches near Ennis, the Hodges retained Hill's services and began an ambitious and impressive building campaign to create a true showplace ranch. Dorothy Hodges, along with Hill, was responsible for selecting all of the building designs, setting, and landscaping. Hill's rustic but modern style is embodied in the main ranch house, two houses for ranch hands, a bunkhouse, a guest cabin, a cookhouse and commissary, three livestock barns, a slaughterhouse, a poultry barn, and miles of board fence. Hill even designed original rustic furniture for the house, bunkhouse, and guest cabin.

The architectural gem at the Jumping Horse is the massive show barn, a ten-thousand-square-foot, cross-gable, gambrel-roof structure. The center of the barn boasts a fifty-foot tall ceiling lighted by a charming barn-shaped, multipaned window in each gable. The entrance doors are twenty feet tall and reportedly weigh seven hundred pounds each. This central portion of the barn was used for livestock shows and sales; it had a balcony on the second floor for viewers. The matching side wings held spacious, well-lighted horse stalls, with room for a dozen horses in each wing.

Central show floor inside the show barn

Horse barn

Remarkably, this elaborate ranch was built primarily during the early years of the Great Depression, thanks to the invention of flash freezing. Wetmore Hodges, a business researcher at Harvard University, became a primary investor in Clarence Birdseye's General Seafood Company in 1924. Hodges and other investors helped Birdseye develop and market his flash-freezing technology, and in 1929 the company was sold to the Postum Corporation, the deal carrying the largest price ever paid for a company at that time.

Their fortune allowed the Hodges to continue with their elaborate plans well into the 1930s, supporting a small army of workers on their ranch at a time when ranchers were letting help go and putting off any building plans. About seventy-five people were employed at the ranch in peak season, helping Wetmore raise Thoroughbred racehorses, Percheron work-horses, Quarter Horses, mules for the U.S. Forest Service and U.S. Army, and over a thousand head of purebred Herefords. To support the staff, other ranch hands tended a herd of dairy cattle, chickens, turkeys, and pigs.

After about ten years with little or no profit, Hodges had to scale back his operation to only cattle and hay. Wetmore Hodges's sons recall their dad's remorse at not being "able to do more" to rebuild the local area economy, and locals remember Hodges often saying that he was doing all he could to provide work in the Madison Valley. After World War II, Hodges was eager to devote his time to developing other business interests, and he sold the ranch in 1954. Since then, the elegant buildings and landscaping have been well preserved by a chain of owners dedicated to carrying forward the Hodges legacy.

Horse barn interior

Bos Farm Barns GALLATIN COUNTY, 1907–1908

John and Rose Brouwer still use all the Bos Farm barns and other outbuildings.

The Bos family farm is a shining example of a diversified farm developed by Dutch immigrants. John Bos Sr. brought his family to Manhattan in one of the first waves of immigrants to arrive from the Netherlands in 1893. Bos ran a blacksmith shop in town for a number of years, then began farming south of town in 1904.

The log granary was already on the property, most likely built by the first owner, Lewis Young, before 1897. Little is known about the log building or its builder, but it is fair to speculate that the two log portions were built separately as a barn and granary and later connected.

As soon as Bos bought the property, he began a building campaign, adding a new residence, blacksmith shop, chicken and brooder house, hog house, and combined horse and dairy barn. According to ledgers carefully kept by Bos, he designed and built the barn in 1907 or 1908 with the help of four hired men. The lead carpenter was paid $228; the others earned $10, $13.50, and $24, respectively. Most of the hardware was made in the blacksmith shop. The uniform design and plan of the buildings suggest that Bos may have consulted barn-plan books or Extension Service bulletins.

The barn was set up to accommodate a variety of animals. The east side of the barn had horse stalls, six stanchions for milking, and a pen for the stallion. The west side had pens for calving. The loft was used to store hay, and there was a large bin for grain that funneled into the dairy cow stalls. The other outbuildings reflect the range of farm-building design of the early twentieth century. The hog house is an especially good example of its type, one that has all but vanished in Montana.

A giant draft horse held all of the Bos children and cousins in this classic 1940s photograph. The barn and granary are in the background.

The log granary is an early Gallatin Valley outbuilding, constructed sometime before 1897.

Hog house

Waterman Barn

GALLATIN COUNTY, CIRCA 1885

A landmark in the Gallatin Gateway area, this barn is distinctive for its saltbox-type roof. Less common in Montana, the saltbox barn is often seen in Dutch settlements in New England, southwest Michigan, and elsewhere. There are two such barns remaining in the New Holland area of the Gallatin Valley. More a building shape than a building style, the saltbox takes its name from the slanted lid on the wooden boxes used to store salt in colonial times. Although some houses and barns with this roof type result from a shed-roof addition to a standard gable-roof structure, the Waterman Barn appears to have been built all at one time with this steeply pitched roof.

The barn's design may have been influenced by the colorful and well-traveled life of its owner and early Bozeman pioneer, Christopher H. Waterman. Waterman was raised on a farm in the 1820s and 1830s in western New York State. At the age of twenty, he spent a year working in Michigan. He then spent much of his twenties and thirties at sea, eventually sailing as captain on a large vessel to Havre, France. At the time the Civil War broke out, Waterman moved with his wife to Saginaw, Michigan, where he was a cooper. It seems very likely that his upbringing and career brought him past colonial-era structures, many with saltbox plans.

Near the end of the Civil War, Waterman moved the family west with a wagon train of about two hundred emigrants, becoming one of the earliest settlers in the Gallatin Valley. He first built a log cabin and later a log barn with local timbers.

The Waterman Barn is one of the few examples of a saltbox-roof barn in the Gallatin Valley.

In about 1876, health problems forced the family back east, but they had returned by 1880 and were working again to enlarge and improve the ranch, amassing over eighteen hundred acres. The barn was most likely constructed in this second period, perhaps with the help of Waterman's sons and when milled lumber for siding was more readily available.

Dyk Barn

GALLATIN COUNTY, CIRCA 1910

The monitor-style roof on the Dyk Barn was popular in the Amsterdam area. The "K" in the gable end represents the Kamp family, owners since 1957.

The first Dutch settlers in the Gallatin Valley came in 1892, encouraged by A. J. Wormser of the Presbyterian Church. While a number of the first settlers left the area reporting harsh conditions for farming, many persevered and encouraged their families and friends to immigrate from Holland and Michigan. Many came in the first years of 1900, when rainfall was favorable and promotion of the West by railroad companies was at its peak.

Sam Dyk came with his parents to Montana in 1903 when he was about twenty. He was married in 1908 and began farming north of Amsterdam. He likely built the barn shortly thereafter to house workhorses and some dairy cattle. Dyk was one of the first potato farmers in the county. After his retirement, he sold the farm to Thomas J. Kamp in 1957. In 1977, Kamp's son John purchased the original house and this barn.

The monitor-roof barn found favor among many Dutch settlers in the Amsterdam area. A study of Amsterdam barns in the late 1960s found that a third of the barns still standing had a monitor roof. Although early Dutch settlers in colonial America built heavy-timbered barns much like those in their mother country, the Dutch barns built in Montana's New Holland area were of a different era, and most favored the standard gambrel or monitor roofs illustrated in such publications of the day as Extension Service bulletins and farm building catalogs rather than following ethnic tradition.

Gowan hayloft

Gowin and Stimson Barns

GALLATIN COUNTY, 1917

The Gowin and Stimson Barns in Gallatin Valley's Springhill community are twins built at the same time by Taylor Brothers Construction in the summer of 1917. Construction crews set up tents at each location and, according to legend, competed to see which crew would finish their barn first. The crew working on the Gowin Barn (yellow with white trim) won the competition, even though the barn was ten feet longer.

These two Gothic arch barns represent the farmer's dream of having an "unobstructed loft, free from interior braces." The Gothic arch roof, developed after 1910, was a response to farmers' needs for a more spacious hayloft. The innovation of balloon and lumber-truss framing in the late 1890s first made possible the common gambrel roof, which offered a more open hayloft space without the obstacle of heavy timbers. The Gothic arch roof took the design a step further, eliminating all interior post supports in favor of having bent or laminated rafters meet at the roof peak. In both the Stimson and Gowin Barns, hay could be stacked floor to ceiling, gable end to gable end, leaving little space for barn cats to play.

The Gowin and Stimson Barns were both built in 1917. They are identical, except that the Gowin Barn (far right) is ten feet longer.

(near right) Stimson barn

Sanders Dairy Barn

GALLATIN COUNTY, CIRCA 1900

The old "Pink Barn," as it is known today, was part of a dairy operation purchased in 1929 by Charles J. and Beulah Sanders. It was used as a dairy barn through the 1970s as the farm was one of two that delivered milk to homes in Belgrade for almost forty years. The barn also housed workhorses and was used to shelter calves during the winter. Large doors at the front and rear left side allowed a team and wagon or tractor to pull in, load up, and exit the other side. Up in the loft, many tons of hay were piled in order to feed the milk cows. A large hayfork with eight claws was used to lift and pile the loose hay and later to stack small square bales.

The barn is easy to identify as a dairy barn because of its companion, the tall wood-stave silo. Silos like this one, constructed of vertical wood boards with circular iron braces, were popular in the United States from the late 1890s to the 1920s, but they never proliferated in Montana since so many farms and ranches raised beef cattle that grazed on the range. Dairy cows, by contrast, required shelter and feed, which was stored in these giant, airtight containers. Wood silos are rare to see these days because so many were abandoned in the 1950s and have since fallen victim to Montana's harsh weather.

Sadly, the Sanders Dairy Barn was demolished in 2010.

Harsh weather does not account for this barn's rosy hues, however. Originally, the Sanders barn was painted a deep barn-red, with a big white star painted in the south gable end. In the early 1980s, some local painters offered to paint it for a bargain price. The deal was too good to be true, and the cheap, watered-down red paint faded to pink in just a few years. Today, several pink barns and outbuildings are still scattered around the valley, making it easy to see where those painters left their mark.

Venhuizen Potato Barn

GALLATIN COUNTY, LATE 1940s

A giant pile of potatoes inside the enormous underground barn

Farmers of Dutch heritage in the Gallatin Valley have been growing potatoes since the late 1800s. Sam Dyk (see page 113) was one of the first potato growers to produce on a large scale, beginning in the early 1900s. Today, there are about twenty-five families growing seed potatoes in the valley, with approximately ten thousand acres in cultivation. Potato farming is vital to the economy of the Gallatin Valley. While some farmers grow only the classic Russett-Burbank, the Venhuizens grow up to fourteen potato varieties.

The Venhuizen Potato Barn near Manhattan was built in the late 1940s and is a good example of a typical potato barn. The low-slung structure is probably the least noticeable barn type as the roof is all that peeks out of these giant underground storage buildings. Early potato barns were essentially just large root cellars built into a hillside, but as the technology of the twentieth century advanced, barns like the Venhuizen Potato Barn were built like a large concrete basement covered with a wide, low metal roof. Most seed potato farmers harvest potatoes sometime in the fall and then wait to sell them until buyers are ready to plant them. Since potatoes should not be frozen, air is circulated around them all winter and through early spring to keep the potatoes cool and fresh.

Jezick Barn

LEWIS AND CLARK COUNTY, CIRCA 1890

This stone barn, just up the hill from downtown Helena, is one of the few "in town" barns left in Helena. It is a wonderful example of the work of Croatian stonemasons who helped build prominent structures in the capital city. The barn, built by Frank Jezick in the late 1880s, once housed dairy cows and was a workshop for his team of stonemasons.

Frank Jezick immigrated to Montana in 1882 and began work on the Hale Reservoir at the top of Clancy Street. He soon found other work on buildings such as the Clark-Parchen mansion, the Montana Club, and the Helena High School (now demolished). Eventually, he was responsible for bringing some thirteen other Croatian men to Helena between 1888 and 1900. Nine of the men were stoneworkers, ranging in age from fifty-six to eighteen, who lived in the one-story building near the barn.

The clipped-gable, or gherkin-head roof, and limestone walls on the barn reflect centuries-old building traditions commonly found in central and southeastern Europe. Luckily, the barn was not severely damaged by Helena's devastating 1935 earthquake, which resulted in many of the city's beautiful masonry buildings, including the Jezick house, being repaired with a coat of stucco.

The Jezick house and barn are located a stone's throw from downtown Helena.

Sheared ewe leaving the shearing shed

Sieben Ranch Barns

LEWIS AND CLARK COUNTY, CIRCA 1950

The sheep barns on the Sieben Ranch, built in the mid-twentieth century, are much like modern-day warehouses, built long, wide, and low in order to house hundreds of sheep during bad weather, lambing, or shearing. They were built with utility in mind. Today, they are historically significant for their association with Henry Sieben and his legendary career raising sheep and cattle in Montana.

Henry Sieben was only seventeen and an orphan when he arrived in Montana in 1864 with his older brother, Leonard. With no money and little education, he first worked as a ranch hand and later ran a freighting business with his brother. Business was slow in the fall of 1867, so they bought a herd of old oxen, fattened them through the winter, and sold them to a butcher for a tidy profit in the spring. This success in the livestock business led them to establish their first ranch on the Smith River in 1870. In 1875, along with another brother, Jacob, the Siebens bought twenty-two hundred head of sheep and worked over the next four years to upgrade their flock. In 1879, the brothers dissolved the partnership, with Jacob taking all the sheep and Henry the horses and cows.

Nearly twenty years later, Henry Sieben and then-partner Thomas Grimes got back into sheep raising, purchasing the expansive Malcolm Clarke place. One of the oldest ranches in the state, the property encompassed a fifteen- by twenty-five-mile area in Helena's Prickly Pear Valley. Sieben and Grimes ran cattle and a small band of sheep for the next fourteen years and, in 1907, bought the Cannon Brothers Ranch near Cascade, adding sixteen thousand acres and 1,735 head of cattle. To this operation, Sieben brought twenty-two hundred head of sheep from Chinook to add to the seven hundred on hand at the Clarke Ranch. The sheds on the Sieben Ranch reflect this rigorous expansion of the business.

(top) The front half of the shearing shed has stalls and equipment for each shearer as well as an open area for wool sorting.

(bottom) Hundreds of ewes give birth to lambs in this long, low lambing shed.

(left) The back two-thirds of the shearing shed is wide open and big enough to hold hundreds of sheep waiting to be sheared.

(above) The handwriting is on the wall inside the shearing shed—including that of U.S. Senator Max Baucus.

Sieben owned one of the largest sheep operations in the state and was an active and honored member of the Montana Woolgrowers Association; in his later years, he was elected honorary lifetime president of the association and was inducted in 1961 into the Hall of Great Westerners at the National Cowboy and Western Heritage Museum.

After Sieben's death in 1937, the ranches were divided among his daughters and sons-in-law. At its peak in the 1960s, the Prickly Pear operation included about five thousand head of sheep. Today, Sieben's great-grandson, John Baucus, manages the ranch, and the sheep sheds are still used for lambing and shearing, although they serve a smaller flock of sheep—about twenty-four hundred head of Rambouillet, the signature breed of Sieben Livestock Company.

Silver King Ranch Barn

LEWIS AND CLARK COUNTY, 1914–1916

The architecture of the Silver King Ranch Barn reflects the able craftsmanship of a builder well versed in rustic log building traditions. Constructed between 1914 and 1916, this gambrel-roof barn is solidly built of round logs with square corner notching; it features log walls that rise to the full height of the peak of the roof structure.

By the early twentieth century, the availability of railroad-shipped materials made it increasingly common to build balloon-frame barns with milled lumber, and the use of peeled logs faded into the past. Modern barns built with laminated two-by-fours or two-by-sixes, instead of large posts and beams, allowed for wide-open haylofts and expanded storage. The Silver King Barn represents a hybrid approach, one that combines early-day log walls and joinery with a spacious hayloft. The log walls extend up to the roofline and are braced with horizontal logs in the roof peak. This approach must have been more labor-intensive and expensive, but it gave the desired aesthetic effect as well as a large, open loft space. The ranch house nearby is also built in a similar rustic style.

The owner, Owen Byrnes, was trained as a mining engineer in the East and was well known locally as the developer of the Jay Gould Mine, one of the largest mines in the Lincoln Mining District. He raised Belgian draft horses here at Silver King for his Lander Valley Ranch Company. For years, Byrnes led hunting pack trips from his ranch through the surrounding national forest and up nearby Silver King Mountain.

The Silver King Ranch, with its rugged log house and barn, survives as a wonderful example of rustic-style log building well suited to the forested setting and to Byrnes's active life as an outfitter, outdoorsman, and four-term state legislator.

When owner Owen Byrnes built this barn on the Silver King Ranch, milled clapboard siding, sawn lumber, and laminated rafter beams were all available, but Byrnes opted for the rustic log style.

Even the hayloft walls and support structure are made entirely of peeled logs.

(above) The rear wall of the barn abuts the hillside and is constructed of dry-laid stone.

(right) The unusual curve in the barn resulted from Auchard's wish to connect his horse barn (background) with his granary (foreground).

Auchard Barn

LEWIS AND CLARK COUNTY, 1887–1888

Following the devastating winter of 1886, when a huge percentage of Montana's cattle herd was lost, ranchers built some of the largest and most impressive barns still standing today in Montana. Cattlemen with means, such as David Auchard in Lewis and Clark County and William Child in Jefferson County, began acquiring and fencing large tracts of pasture land and building large cattle barns to store winter feed and shelter cattle during bad weather.

The Auchard Barn stands out because of its unusual curved central section (seen in both photos to the right), built to connect two earlier barns: the small 1870s log granary on the south and the early 1880s gable-roof horse barn on the north. This was no easy feat, and the challenges were compounded when Auchard's contractor quit before the job was finished. Luckily, Auchard's brother Richard was able to complete the task.

In addition to the tricky curve, the barn is a remarkable structure, built with carefully stacked and dry-laid stonewalls, finely hewn timbers, and masterful mortise-and-tenon joinery. Broadaxes and adzes have been unearthed on the property, a reminder today that these giant barns were raised without the aid of power tools—just the strength of horses, men's arms and backs, and sharp iron. In addition, Auchard's barn is a bank barn; the rear portion of the barn is built into a slight bench in the landscape, allowing hay wagons to pull right up to the rear hay door for loading. On the interior, the entire rear wall is constructed of dry-laid stone.

When the barn was completed, it served a variety of uses on the ranch. Between the north-end stable and the south-end granary, the long central section had feed stalls for cattle and a large open pen that could house hundreds of cattle during a storm. Large doors in the midsection allowed hay wagons or carriages to pull directly into the barn, and the upper story was likely large enough to store five hundred tons of hay. The loft, loaded to full capacity, would have been a sight to see.

David Auchard was an early Montana pioneer and one of the most successful late-nineteenth-century ranchers in Lewis and Clark County. After the days of the open range were over, he amassed about thirty thousand acres in the Dearborn countryside, buying up railroad lands and adjacent homesteads as they became available. When he died in 1902, he left a large bequest to the Masonic Grand Lodge in Helena, which eventually led to the building of the Masonic Home of Helena.

In 1910, the Burggraff family bought the house and barn from the Auchard estate. It has been in the Burggraff family since then.

Milford Colony Sheep Barn

LEWIS AND CLARK COUNTY, 1947–1948

(left to right) Paul Kleinsasser, John Kleinsasser, and Ed Hofer

This enormous sheep barn on the Milford Hutterite Colony near Wolf Creek was built in late 1947 and early 1948 by members of the Milford Colony. Its style is unique in that it appears to have no first story, or that the gambrel roof was lifted off and put on a concrete foundation. Indeed, the colony intentionally built the barn this way to house hundreds of sheep during lambing and shearing and to store many tons of hay and feed in the hayloft.

The Milford Hutterite Colony was established in 1945, splitting off from a colony in Alberta that had grown too large. Hutterites, a communal branch of Anabaptists, who, like the Amish and the Mennonites, trace their roots to the radical Protestant reformation of the sixteenth century, live in rural areas and historically have maintained self-sustaining farms and ranches. There are fifty colonies now in Montana.

By the 1940s, lightweight truss systems allowed barns to be built to almost any length and very generous widths, with no interior posts or piers to block the flow of animals or people. Between the trusses in the hayloft, many of the lambers left their marks on the walls, recording their initials, dates, and the number of lambs born.

Today, the old sheep barn is used for storage, a reflection of changing times for the Hutterites. While they will continue to live communally, practice pacifism, and speak German, they no longer raise sheep but instead have added manufacturing stone pavers for landscaping to their diversified farming operations.

Meyer/Staubach Creek Ranch Barns

BROADWATER COUNTY, 1925

Antone Meyer's round-roof barn in Broadwater County is a good example of modern roof construction on top of traditional log walls. By 1925, many farmers were attracted to the round-roof shape because it provided even more loft space than a gambrel roof, and compared with the old method of timber framing, it was relatively easy to put in place. But Meyer did not take the easiest route, ordering all the lumber precut and delivered to the ranch. Instead, he chose to take advantage of the nearby forests to build the barn walls. He ordered only the roof rafters from out of state.

Completing such a large round-roof log barn must have been a source of pride and great satisfaction for Antone Meyer, who upon completion of his barn carved his name into one of the rafter beams. Meyer was justified in his pride, given that it must have taken many man-hours to cut the timber from the Elkhorn Mountains; haul it to the site; then peel, notch, and stack the logs just so. And it must have taken a team of men and horses to lift the laminated rafter arches into place and put a roof over the whole thing. Building a barn, even a small one, while tending to animals and family and finances represents such a herculean effort that it is surprising that more barns do not bear the marks of their builders.

The handsome brick house and many of the secondary buildings on the Staubach Creek Ranch were put up by Valentine Staubach. In 1866, Staubach, along with his wife and six children, were the first settlers on Beaver Creek near Winston. They came with four hundred other men and women on one of the first large wagon trains into Montana Territory. As early settlers, they were squatters on unsurveyed government land, but over the next forty years they improved and developed the land into a successful cattle ranch of fourteen hundred acres.

(above left) Antone Meyer built the round-roof barn; the gable-roof barn was brought in from Canyon Ferry.

W. C. Child Ranch Barn

JEFFERSON COUNTY, 1887

Child named his ranch White Face Farm in honor of the white-faced Herefords he raised. Only six years after the barn was built, Child passed away, and the barn sat mostly vacant for fifty years. When Paul Kleffner bought it in 1943, all seventy windows on the barn had been broken. Over the course of thirty years, he repaired them while returning the barn to its 1887 grandeur.

Successful Helena mining man W. C. Child began ranching in 1885. He established the White Face Farm east of the city, stocking it with some of the first purebred Herefords in Montana Territory. Then came the devastating hard winter of 1886–1887, when deep snows and months of freezing temperatures decimated entire herds of cattle. When spring arrived, Child built the massive stone barn to house and protect his stock from another vicious winter. This monumental three-story barn with full-length side wings made of local fieldstone measures one hundred feet by one hundred feet on the first two levels and one hundred feet by sixty-five feet on the third hayloft level. It was designed to hold five hundred cattle; 350 tons of hay could be stored in the loft. Enriched by his mining enterprises, Child did everything in grand scale and

KLEFFNER RANCH
ESTABLISHED 1889 BY W.C. CHILD

later built an octagon-shaped stone house nearby, just for meetings and parties.

After Child's death in 1893, the property passed from owner to owner, its buildings deteriorating over the years. By 1943, when the Paul Kleffner family moved to the ranch, all seventy windows on the barn were broken out, the wood siding was weathered and dilapidated, and four feet of manure was piled on the lower-level floor. Over the next thirty years, Kleffner carefully restored the barn, spending approximately thirty thousand dollars during that time to return it to its 1887 appearance. The building earned a National Trust for Historic Preservation Barn Again! Merit Award in 1988. Today, it is not only a beloved landmark in the East Helena Valley but also one of the most desirable destinations in the region for parties, weddings, and family reunions.

Marks Barn

JEFFERSON COUNTY, 1923

The Marks Barn near Clancy was built in 1923 by Merle Marks with logs cut near Warm Springs Creek, some measuring forty feet in length. Marks used cottonwood on the interior, a more durable wood than pine and suited to shouldering the weight and wear and tear of the large Percheron draft horses that Marks raised. Although most Gothic-roof barns were built with laminated rafter trusses to provide for an unobstructed hayloft floor, the Marks Barn also has hand-hewn posts and beams for an added measure of support. The high-profile roof on this barn is also embellished with the family brand, the Cross Half-circle.

The barn, the brand, and the surrounding lands have been in the Marks family since the 1880s, passed down to each new generation and carefully preserved. When Ernest Marks first came to Jefferson County from Maine, he was a freighter in the days when the local mines were booming. Later, when the mines played out, he turned to cattle ranching. In the 1920s, his son, Merle, continued raising cattle and also draft horses, building this barn to house dairy cows and workhorses. He also built a sawmill nearby to augment ranching income. The family business was next passed to Merle's son, Bob, who continued his father's and grandfather's businesses—ranching, logging, milling, and construction. He also served in the Montana House of Representatives for twenty years. Today, the Marks family still uses the barn for horses, and both Marks-Miller Post and Pole and Marks Lumber are mainstay businesses on the edge of Clancy.

Litening Ranch Barn

BROADWATER COUNTY, 1901

The barn with the "Litening" gasoline advertisement on top was a familiar landmark along U.S. Highway 287 for nearly thirty years, until the owners were forced to reroof.

The Litening Barn was erected in 1901 by John Walter, a professional barn builder from Missouri who hired on to build the barn while he was in Montana visiting family. With its classic gable roof and tall, elegant ventilator, the barn was distinctive even before the word "Litening" first appeared on the roof in big white letters. Until the present owners were forced to reroof, it was a familiar sight for more than thirty years along Highway 287. The simple center-aisle barn has horse stalls on each side and a tack room near the center. The long length of the building allowed for ample hay storage despite a relatively low-pitched gable roof over the loft.

Litening brand gasoline was advertised on signs, billboards, buttons, and road maps across the western states by the Yale Oil Company of Billings, primarily during the 1930s and 1940s. Although barns as billboards are not a common sight in Montana, many eastern farmers—and travelers—fondly remember the Mail Pouch Tobacco ads painted on roadside barns from 1890 to 1992. At the height of the program in the early 1960s, there were about twenty thousand Mail Pouch barns spread across twenty-two states.

Sologub Dairy Barn

JEFFERSON COUNTY, 1950s

Elk Park was historically the main dairy-producing area serving the city of Butte, and today there are several farms and barns that reflect that heritage. Traveling along Interstate 15 between Boulder and Butte, drivers cannot miss the big white barn with bright green trim in Elk Park. Joseph Peter Sologub, the eldest son of Russian immigrants, began a dairy with his mother and two younger brothers in the early 1920s. Upon the death of the youngest brother in the 1930s, Joseph bought out his mother and brother. Over the years, he developed his operation into one of the largest in the area, and by 1950 he was ready to replace the old log barn. When it came time to build the new state-of-the-art dairy barn, Joseph, with the help of a couple of miners, constructed the new barn around the old log barn so that dairying could continue uninterrupted. At the peak of operations in the 1960s, Sologub milked up to eighty-five cows at one time. The barn, with its spacious interior, long rows of ribbon windows on each side, and modern milking equipment, was likely a model for area farmers, most of whom were still using much smaller log or wood-frame dairy barns.

Unusual zigzag bracing reinforces the rafter trusses on the interior of the dairy barn roof.

Francone and Parini Barns

JEFFERSON COUNTY, 1920s

The opening of the copper mines on Butte Hill in 1876 gave rise to the cities of Butte and Anaconda and drew thousands of hopeful immigrants to the communities. Beginning in the late 1890s, a number of Swiss-Italian families settled in the Meaderville district on the outskirts of Butte, and most of those men started out working in the mines. But, finding the outlying high-elevation landscape to be much like that in the Alps, some families continued with tradition and began dairy farming in the nearby Elk Park Valley in the early 1900s.

The Francone and Parini families both settled along Elk Park Road. In the 1920s, George Francone and his brother William built a long and spacious red dairy barn to house fifty cows. George chose a typical barn design with a gambrel roof and red siding. The hayloft floor and roof are supported by logs cut in the Deer Lodge National Forest nearby, but much of the milled lumber came from lumberyards in Butte. Francone's ability to buy and haul lumber from Butte suggests that his dairy business was prospering at the time.

Despite the availability of milled lumber and standard barn plans in the 1920s, many farmers still were building log barns, mostly out of necessity in the lean years of the late 1920s and throughout the Great Depression. In 1929, when Rudy Parini was only eight years old, he helped his older brother, John, build a new log dairy barn with a gambrel roof to complement the one their father had built many years before. From the rustic saddle-notched corners to the simple batten doors, the barn has all the marks of a structure made without standard, store-bought materials. The resourceful brothers even made their own willow brooms for cleaning manure out of the waste gutters.

The Parini barns remain in good condition today, even though they had to be moved from their original location farther to the west when Interstate 15 was constructed in the 1960s.

(opposite) Francone dairy barn

(above left) Esther Francone

(above right) The smaller barn on the right is the original barn built by Parini.

Carlson Hay Barn

JEFFERSON COUNTY, 1900

This barn is a quintessential hay barn, built to store the feed essential to keeping livestock through long Montana winters. Unlike more refined horse and dairy barns that dot the landscape, hay barns with their unfinished rough-hewn siding and simple dimensions have never enjoyed the "curb appeal" that many other barns possess. Built when the town of Bernice was brimming with newcomers seeking their fortunes in the mines, the barn stored hay for use at the dairy farm and at the stage station at Bernice. Other hay barns in the area served a dual use—the barns on Torpey Gulch and Moonshine Gulch both housed homemade stills.

Geary Ranch Barns POWELL COUNTY, CIRCA 1880

The barns on the Geary Ranch are among the first buildings constructed in the Helmville area and are wonderful examples of mountain horse barns built by Irish immigrants. Built in the 1870s or early 1880s by John Geary and his cousins James and Michael Geary, these barns exemplify common early barn characteristics with their rough log walls, irregular corners, and lack of any eye-catching architectural details. Modest and often overlooked, the barns are important reminders of the work of Montana's earliest ranchers, whose toil was a key factor in developing the territory's livestock industry.

The Gearys first visited what was to become Helmville in 1866 and had taken up land on Nevada Creek by 1867. All three men lived together in a one-room cabin and spent many long hours improving their ranchland. It is not known which barn was built first, but it appears that each started out with a large, central log room with stalls and a hayloft above. The side wings were added on as time permitted. Descendants of the first Gearys say that, inside the horse barns, they often find the remnants of old beer and whiskey bottles buried not too far below the surface of the dirt floors.

By 1870, the cousins had established the 3-G brand and held excellent water rights along Nevada Creek that aided in raising their Shorthorn cattle. Eventually, James and Michael Geary married and moved on to develop their own ranches. John Geary stayed on the home place and married Marianne McCormick in 1885. Many third-, fourth-, and fifth-generation Gearys still live in the area, and John Geary's great-grandson still uses these barns in his ranching operation.

The humble, rough-hewn log buildings on the Geary Ranch are excellent examples of the early outbuildings constructed by most who settled in the Helmville area in the late 1870s and early 1880s.

Every year, the Grant-Kohrs Ranch National Historic Site puts on demonstrations of haystacking with one of Montana's own beaverslide haystackers.

In this 1885 lithograph, the Kohrs livestock take center stage.

Grant-Kohrs Ranch Barns

POWELL COUNTY, CIRCA 1870s

The Grant-Kohrs Ranch could be called the cradle of Montana stock raising. Even before the gold rushes at Bannack and Virginia City, there were cattle ranchers in the Deer Lodge Valley. Some of the first cattle were brought here by Johnny Grant, who began wintering cattle in western Montana valleys in the 1850s. In 1866, Conrad Kohrs purchased the Grant ranch, which included a stately home, barns, land, and 365 head of cattle. He formed a powerful partnership with his younger half-brother, John Bielenberg; they came to control about 1 million acres in eastern Montana, and their cattle grazed on 10 million acres of public land in four states and Canada. At the home ranch, they became known for raising registered Shorthorn and Hereford cattle as well as Thoroughbred and draft horses.

While Kohrs spent his time on the business end of ranching, it was Bielenberg who was "out in the field," and his contributions to the livestock industry were many. Among them was his development of Big Circle horses. By crossing Thoroughbred studs to hardy native mares, Bielenberg bred cow ponies that could do a twenty-mile circle in half a day during the roundup. This was a huge territory to cover in trailless, broken country, where cattle might scatter over 2 million acres within a single grazing district. These forerunners of today's Quarter Horses were in high demand throughout the territory. Bielenberg occasionally raced his Thoroughbreds

During Conrad Kohrs and John Bielenberg's days on the ranch, each stallion had its own barn. These simple log buildings are some of the oldest on the property.

against those of other prominent breeders, including copper king Marcus Daly.

The stallion and bull barns remaining on the ranch today represent Kohrs and Bielenberg's contribution to stock raising in Montana. Conrad Warren, Kohrs's grandson, remembered that "There were four or five stallions in the little barns scattered around the flat below the house. . . . Each barn had a hayloft and a feed box of oats in a little feed room with a ladder going up to the loft. [Ranch hands] didn't have to go in with the stallions to feed and water, a safety precaution."

While the Grant-Kohrs Ranch has many fine buildings and outbuildings, the stallion barns are among the oldest on the property and stand remarkably unchanged since the 1870s and 1880s, when Bielenberg developed the ranch's Thoroughbred horse operation.

Current owner Bill Geary comes from a long line of Nevada Valley ranchers, dating back to the 1860s.

Moore Ranch Barn

POWELL COUNTY, 1906

Although it is common to see beaverslide haystackers in Blackfoot Valley hay fields, it is not so common to see a haystacker built inside the loft of a barn. The complex gambrel and gabled roof on the Moore Ranch Barn houses an ingenious system for bringing hay into this giant hayloft. Loose or baled hay was loaded through the drop-down door from a wagon below and into a hay basket, then pulled by rope and pulley up the rack and into the barn by a team of horses. Hay could then be dropped down a large chute and distributed to the horse and cattle stalls below. When the loft was not full of hay, the cavernous interior was large enough to accommodate a full basketball court.

This large, multipurpose barn on the Moore Ranch near Helmville was built around 1906 by John Moore's son-in-law, David Raymond, an English-Canadian carpenter from New Brunswick. Raymond was well known in the Nevada Valley as a builder and is said to have had a hand in building half the houses and barns in the area. He was a resourceful man, and the story goes that when he arrived in Montana in 1900, he rode a bike from Missoula to Helmville as he had no other way to get there.

The stepped-roof extension houses a beaverslide haystacker used to get hay from the wagon into the loft.

John W. Blair's legacy is still very much alive as The Company Ranch is still part of a thriving beef cattle operation.

(right) Horse barn

The Company Ranch Barns

POWELL COUNTY, CIRCA 1895

Sheep barn

(opposite) Unpeeled log buttresses help support the hayloft walls in The Company Ranch horse barn.

The crisp red barns on The Company Ranch represent the prosperous career of Helmville banker and rancher John W. Blair. Arriving in the Helmville Valley in 1872, just as a number of other early settlers began farming and ranching, Blair developed a truck garden and sold fresh vegetables to area miners. After inheriting money in the early 1890s, he greatly expanded his humble farm into a full-fledged cattle and sheep operation, eventually amassing over twenty-two thousand acres. By the early 1900s, Blair owned more land than any other taxpayer in the county and had one of the largest ranches in the state.

The large monitor-roof horse barn, low-pitched gable-roof sheep barn, and grain elevator (not pictured) were all likely built in the late 1890s. The horse barn features a central drive-through alley with stalls along each side and side sheds holding more stalls and room for storage. It easily accommodated a dozen workhorses as well as feed, tack, wagons, and other equipment.

The large sheep barn is a good example of a typical sheep shed, with its broad gable roof, large open plan, and doors on each side for driving sheep in and out during shearing. On the sheep barn's interior walls, sheepshearers marked their annual tallies—a workingman's record of life on the ranch.

CHAPTER 5

Barns of the Missouri River Country

THE MISSOURI RIVER runs like an umbilical cord across Montana's broad plains, flowing more than seven hundred miles from its headwaters near Three Forks to the North Dakota border. From there, the Big Muddy stretches another sixteen hundred miles to meet the Mississippi at St. Louis, making it the longest river on the continent. It gives life to farms and ranches all along the way.

Meriwether Lewis and William Clark recorded the attributes of this region on their journey, and fur traders followed. Lewis was awed by the buffalo that roamed the Sun River Valley in 1806. "I sincerely beleif," he wrote, "that there were not less than 10 thousand buffaloe within a circle of 2 miles."[1]

A chain of forts soon lined Montana's Yellowstone and Missouri rivers, running from Fort Union (built in 1828), at the rivers' confluence, to Fort Benton (built in 1846), at the head of navigation near the Great Falls of the Missouri.[2] Trading posts and fortifications ushered in the era of fur trade and steamboats on the upper Missouri and changed the face of the plains forever, as millions of buffalo and beaver gave way in the decades that followed to hundreds of thousands of cattle and sheep.

Whether military or civilian owned, many early forts utilized adobe brick construction, with corner bastions and log-walled stockades for security. Fort Union, Fort Benton, and Fort Owen all had such buildings. Inside the walls, there were living quarters, storage buildings, stables, and blacksmith shops. The earliest barns on record include a mid-1830s dairy at Fort Union and an 1870s stable at Fort Peck.[3]

Fort Benton was a small but bustling outpost from which the gold, hides, and pelts of the territory flowed downstream to St. Louis and world markets, while up the river came guns, whiskey, and Indian trade goods. Government contracts to provision the early military forts and Indian agencies enriched such early traders as T. C. Power and I. G. Baker, who built financial empires by providing these outposts with beef rations, arms, and alcohol. During this era, agriculture took hold in proximity to forts, trading posts, Indian agencies, and gold camps. While barns were a rarity, these frontier enterprises opened the way to farming and livestock operations in the region.

(opposite) Hodson Barn, Cascade County, built 1904

A lion's share of the merchandise that reached Montana in the 1860s–1870s came on steamboats to Fort Benton. Photographed here in 1878, it was the head of transportation on the upper Missouri River. Fort Benton's frontier enterprises opened the way to farming and livestock operations in the region.

The Sun River Valley

Freight roads connected Fort Benton and nearby Carroll to the gold camps of Helena and Virginia City and to the Mullan Road. During the early 1870s, stock raisers from the Deer Lodge Valley drove large herds of cattle into new ranges in this region—through the Helena Valley, north to the Great Falls, and beyond to the plains and inland mountain ranges that frame Montana's northeastern horizons.

The enormous Sun River Valley—flat, open, and well watered—lay along the trail between Fort Benton and Helena and was known to freighters and travelers for its good grasses. The first Blackfeet Agency and Fort Shaw were established here along the Sun River in 1858 and 1867, respectively. Then stock raising took off after Robert S. Ford, a bullwhacker for the Diamond R Transportation Company, drove seven hundred cattle from Denver to the Sun River in 1871 and returned the following year with twelve hundred more. Other freighters took up residence, and cattle barons Conrad Kohrs and Daniel Floweree drove herds in from Texas and Oregon and turned then out onto this new range, sparking an open-range livestock boom.[4]

As their stock foraged freely on the public domain, these Sun River ranchers built substantial houses and outbuildings. Among them was a landmark sandstone barn built by J. C. Adams that stands out in Montana's architectural history. Like many ranchers of the valley, Adams capitalized his livestock business by selling beef and stock to nearby Fort Shaw, and the 1885 J. C. Adams Stone Barn reflects his entrepreneurial success. Built by two stonecutters for a reported ten thousand dollars, the barn was declared by the *Sun River Press* to be "the Marvel of Montana." The paper rightly predicted that "it would stand the storms and decay of a century at least" (see page 164).

Nearby, Captain Thomas Couch, general manager for the smelter at Great Falls, built a massive timber-framed barn in 1890. Couch experimented widely with various breeds of livestock to improve the bloodlines of Montana cattle. The substantial Couch and Adams barns reflect the burgeoning Montana livestock economy during the latter 1800s and the important expansion of new ranges within the Missouri River drainage. The architecture reflects well-honed late-nineteenth-century traditional building methods, including impressively hand-cut sandstone and hand-hewn mortised-and-tenoned timber framing.

The Judith and the Musselshell

After moving livestock through the portal of the Sun River, enterprising ranchers and powerful territorial capitalists spread into the Smith River Valley and then north and east, stitching enormous grassy squares into the quilt of central Montana. By the mid-1870s, cattlemen pushed through to the east, and leaders like Robert Coburn and Conrad Kohrs and John Bielenberg settled large spreads in the rolling valleys of the upper Musselshell Valley. Others went on to the Judith Basin, where deep, loamy soils and protein-rich prairie grasses awaited the first plainsmen to ride herd over these ranges. Granville Stuart remarked in 1879 on the "magnificent bunch grass" but found the Judith "too open and exposed for a cattle range," declaring it instead to be "splendid farming country."[5] Still, in 1879, territorial governor Samuel T. Hauser, A. J. Davis, and Stuart established one of Montana's largest open-range spreads near Fort McGinnis in central Montana, under the DHS brand.

Sheep soon followed, trailed in by T. C. Power, James Fergus, and others. During the late 1870s, enormous flocks ranged where cattle did not go, and by 1879 there were fifteen thousand head of sheep on the Smith River and a staggering number of sheep, sixty thousand head, grazing along the Musselshell.[6] In 1884, Englishman Charles Alston Messiter lamented that the Judith Basin he had known as a hunter's paradise in 1878 was now "a mass of sheep ranches, and not at all improved by the change, a sheep-ranche being an unpleasant place to stop at, the immense number of sheep tainting the whole air."[7] The trend was reinforced a few years later, after sheep proved their resiliency during the hard winter of 1886–1887. Many cattlemen subsequently diversified into sheep raising to hedge against cattle losses, and at the turn of the century, Montana's wool growers topped the nation with some 6 million sheep.[8]

Foremost among them was sheepman Charles M. Bair, who bought a ranch near Lavina in 1891 and by 1895 had forty thousand sheep on ranges near Billings. Three years later, Bair sold out and headed north to increase his fortune mining in the Klondike. Upon his return, he invested his earnings in a ranch near Martinsdale and leased range on the Crow Reservation. By 1910, the Bair Ranch was reported to be the world's largest sheep operation, with three hundred thousand head. That year alone, Bair shipped a trainload of wool—1.5 million pounds—to Boston markets.

Sheep barns differed from horse and cattle barns, affording space under one roof or in adjacent sheds for lambing, counting, dipping, and shearing. Sheep sheds grew to enormous proportions, their designs derived from sheep-raising cultures in Europe and Australia and evolving industrial technologies. The barns associated with the Bair Ranch represent different epochs in the family's history. From the early barns on his foundation ranch to the huge sheds still used at the Bair Museum today, Charlie Bair's log and frame buildings are a treasure. They contain a wealth of history within their walls (see page 192).

In the first few years of the 1900s, expanding irrigation, along with rail promotion and homesteading, increased the tide of settlement across Montana. The "Golden Triangle" prairie region surrounding Great Falls opened up as large irrigation projects harnessed the Missouri River and brought more land into farm production. These conditions also made possible the growth of the dairy industry, and large dairies

Freshly shorn sheep stand near one of the sheep sheds on the Bair Ranch.

with impressive barns rose from the landscape wherever town populations grew. The Ayrshire Dairy and others of this era featured up-to-date designs, breeding programs, and applied dairying technologies. Beginning in 1906, Scotsman Harry B. Mitchell imported Ayrshire Highlander cows from his home country and built wood-and-brick barns to house them. Mitchell's hardy Ayrshire cows stood up to Montana's tough climate, and the barns stood the test of time, as the dairy operations spanned eighty years (see page 174).

Ethnic settlement also shaped the agricultural traditions of the upper Missouri region. Immigrant farmers broke sod, cleared fields, and built in the manner of their forebears. Woven together by homesteading, industry, and railroading, the fabric of Montana homesteading society was a cloth of many nations, and the state's important ethnic settlements and buildings reflect those patterns. Near the Highwood Mountains, the settlement of Korpivaara is today one of the most intact rural Finnish farmstead settlements in the nation. Drawn to work in the coal mines of Belt, a number of Finnish immigrant families took up homesteads around 1900 on Little Belt Creek. For four decades, the Finns here built with log in the finest Suomi traditions—houses with weaving rooms, saunas, and finely hewn and joined barns for horses, chickens, hogs, and cattle. Today, the Heikkila-Mattila and Lewis-Nevala homesteads showcase the purest examples in Montana of Finnish architecture (see pages 170 and 168).

The High Line

Across northern Montana, there is a vast and open stretch of high plains called the High Line. The High Line gets its name from the rail line built across it from the North Dakota border to the mountains of Glacier National Park. This oceanic grassland was once filled with buffalo and was home to the Assiniboine, the Gros Ventre, the Chippewa-Crees, the Blackfeet, and the Lakota. The Missouri, Milk, and Marias

Army engineers designed this brick monitor-roofed cavalry barn at Fort Assinniboine, southwest of Havre, to hold a multitude of military horses for working and patrolling.

rivers braid their way across these plains; they were harnessed in the twentieth century for irrigation, and today they nourish fields of wheat and barley that roll away as far as the eye can see in a landscape filled with grain and cattle from one horizon to the other.

This country was once Indian territory, and the Blackfeet Nation and the tribes served by the Milk River Agency fought to hold on to their lands, even as the buffalo were being decimated. It was after the Battle of the Bears Paw in 1877, the discovery of gold in the Sweet Grass Hills that same year, the establishment of Fort Assinniboine in 1879, and the surrender of Sitting Bull in 1881 that a permanent military presence in the region ushered in agriculture and the first ranchers. From that era, the brick barns remaining at Fort Assinniboine were designed to hold a multitude of horses for working and patrolling. The brick, monitor-roofed cavalry barns and the gable-roofed bull barn, reinforced on the outside with flying buttresses, were designed by army engineers and were lined up across the fort with military precision (see page 202).

As one scholar noted in his history of Montana, "The possibilities of the level well-grassed plains north of the Missouri as a cattle range had not escaped the attention of the earliest travelers."[9] For every river there were cattle ranchers who turned thousands of head into the empty High Line country north of the Missouri, beginning with Thomas O'Hanlon on the Milk River range near the Fort Belknap Indian Agency in 1878. O'Hanlon was followed by some of the biggest names in the territory, and in 1882 the Marias drew river traders Charles Conrad and I. G. Baker of Fort Benton. The Pepin-Broadwater Ranch was also established in 1882, and Conrad Kohrs and John Bielenberg, whose Pioneer Cattle Company seemed to follow Montana's emerging frontiers, came along in 1886. By the mid-1880s, "There seemed to be cattle not only on a thousand hills, but a thousand cattle on each hill." And, as Granville Stuart noted, the country had gone from "being practically uninhabited to one where small ranches could be found along all streams and where settlements were springing up all over the range."[10]

These were the days of open-range grazing, when fences were unknown and the only mark of ownership was the brand on the hip of rangy cows and unbroken horses. The open-range system came to a screeching halt after the hard winter of 1886–1887. Cattle companies blamed the weather, but historians agree that overgrazing had a lot to do with the collapse of the ranges that long, cold season when more than 300,000 head of cattle, 60 percent of the total in the territory, were lost. It brought to the High Line what Merrill Burlingame called "the final stand of the open range." Outfits pushed north across the Missouri River following the reduction in size of Montana's Blackfeet and Fort Peck reservations in 1888. With the roundup of 1889, the big ranches near the mouth of the Judith River crossed the river, looking to spread out to new ranges. Sheep were a major part of this equation; the J. B. Long and the Bielenberg and Toomey companies expanded their flocks to enormous numbers, over 160,000 and 130,000 head, respectively.[11]

Completion of the Great Northern Railway from Minot to Great Falls in 1887 was an intensive project, employing nine thousand men and seven thousand horses. The Great Northern also fueled farming settlement on the High Line, and the first few homesteaders trickled in through the 1890s. The following decade, the trickle turned to a stream as railroad promotion and the general westward movement enticed people to follow their dreams out onto the Great Plains of the continent. The people on the Fort Peck Reservation also launched farming efforts; by 1905 they had "planted large crops and a great deal of grain," although they had to fight for the water rights to sustain their crops.[12]

Across the High Line, farmers plowed a patchwork of land into long furrows divided from one another by small streams

While some settlers to Montana built barns from rail-shipped kits or pattern books, French settler John Etchart built his house and barn out of stone (see page 210), a plentiful local material and a building tradition native to his Basque country.

and irrigation ditches lined with cottonwood trees. Some of the barns they built lasted a century and more, and hundreds of them are still in use, although many beg for a coat of paint or new roof shingles.

While the farm families shaped the land to agricultural use, the railroad itself shaped the patterns of those who settled there. During the first three months of 1910 alone, the Great Northern hauled over a thousand emigrant railcars into northern Montana, substantially boosting the number of farms and farmers. Lotteries for reservation lands, such as those at Fort Peck in 1914, threw more acreage open for settlement. High Line population figures for this period tell the story: 13,300 people in 1900; 26,200 in 1910; and 57,600 in 1920.[13]

From Havre, where James J. Hill, president of the Great Northern Railway, had dispatched a crew of two hundred to help build the new community, to stops along the line where construction crews rode railcars stocked with lumber and hired out to build homestead improvements, building boomed across the High Line.

The homesteaders were a diverse lot. Most were young twenty- and thirty-somethings, and while some had farming experience, many did not, though many brought their work ethic, stamina, and ingenuity to bear. Many were also immigrants, recruited to Montana by friends or by railroad representatives. Again, the numbers are impressive—there were over twenty nationalities counted in Montana in 1910.[14] Occasionally, whole colonies from one locality crossed the ocean and settled together in places like Daneville, North Dakota, and Dagmar, Montana, which was named for the queen of Denmark.

As a result of the melting-pot effect out on the prairie and the varying availability of milled and rail-shipped lumber, the barns in this region are a mix of style and tradition. Danes, Norwegians, and other Europeans all left their mark. At the tiny hamlet of Tampico, French settler John Etchart built a stone house and barn in the tradition of the Basque country (see page 210). Because the region's agriculture developed during the twentieth century, many barns followed pattern books and were constructed with balloon framing and milled lumber. Gambrel, gable, Gothic, and monitor roofs are all to be found. Some barns were built by carpenters drawn to the area; some, by crews who traveled the rails with boxcars full of building materials and ready-to-go plans.

The High Line became a highly productive region for many cereal crops between 1908 and 1917. Despite the drought years of the early 1920s, bumper grain crops—some 2,750,000 bushels per year by the mid-1920s—placed Scobey as the top wheat-loading center in the United States.

But nothing rivaled the vision of the U.S. Bureau of Reclamation when it came to transforming the High Plains. Bureau officials dreamt of a dam that would tame the river and harness its power, and the Fort Peck Dam, begun in 1933 during an era of big federal water projects, was—and remains—one of the largest man-made structures anywhere. When completed, the dam was 4 miles wide, the reservoir was 189 miles long, and more than one hundred farms and ranches lay at the bottom of the lake. Fort Peck stabilized the flows of the Missouri and impounded water for power production and irrigation. In the years that followed, the High Line became known for its abundance of wheat and for growing some of the world's best safflower seed and sugar beets.

The Jefferson Dredge Barn near Glasgow (see page 208) has one of the most interesting Montana barn histories. Completed in 1941 by Karl Reddig and his sons, it was built using materials from the Jefferson Dredge, one of four dredges decommissioned following completion of the Fort Peck Dam. Reddig purchased the dredge at auction for $251, floated it down the Missouri River, then salvaged the lumber, doors, and windows to construct a gambrel-roofed horse barn that still bears the marks of the dredge.

In 1941, Karl Reddig and his sons built their barn (see page 208) near Glasgow using materials salvaged from the Jefferson Dredge (above), one of four dredges decommissioned after construction of Fort Peck Dam.

Erickson Barn

PONDERA COUNTY, 1917

Milking stanchions are still in place in the Erickson Barn.

Hans Erickson, his brothers, and neighbors built this perfect example of a Gothic arch barn and matching chicken coop in 1917. Though the barn is not large, the handsome patina of the one-inch-by-four-inch laminated and bent rafters that support the roof gives the barn's hayloft a cathedral-like quality. Construction photos show the bentwood roof near completion and also indicate that the original look of the barn has changed very little since 1917.

Hans Erickson came to Pondera County from Little Falls, Minnesota, as a young bachelor with his brothers, sister, and parents as well as a large group of their Norwegian neighbors. Between 1909 and 1910, they settled close together on 160-acre parcels, and the enclave became known as the Erickson community. Most farms raised wheat and later barley, but during the dry years after 1917, many who came found it hard to survive on 160 acres. Some families picked up and left, and those who stayed bought up abandoned land or took up other jobs to make ends meet. The Ericksons supplemented their income with carpentry work, building barns in the surrounding counties and commercial buildings in Great Falls and nearby Conrad. The brothers became well known for building "bentwood" barns throughout the area of north-central Montana.

(top) Hans Erickson and neighbors pose for a construction photo showing the skeleton of their signature bentwood round roof.

(bottom) Detail of bentwood rafters

The circular feeding alley contains troughs, on right, that caught the hay dropped down from the loft above.

Fitzgerald Round Barn

TETON COUNTY, 1913

According to Norma Denny, the current owner of the property and granddaughter of Edward Fitzgerald, when her grandfather "had Arthur 'Big' Elliott build the round barn 3½ miles northwest of Collins in 1913, it wasn't to keep Satan from hiding in the corners as some people thought. What Fitz envisioned was a solid structure that would withstand the constant prairie wind with room for maneuvering his equipment and animals and one that was an easily accessible shelter for his draft horses and Dutch-belt cattle."

This 1913 view shows the barn when it was brand-new.

The barn that Fitz built still stands strong and still catches one's eye when driving the quiet road just west of Interstate 15. The barn floor plan is "basilican," with a center alleyway and a side area for stalls. It is sixty feet in diameter, and the roof reaches forty-four feet high. The second story has an open cupola so that air can flow through the haymow. Originally, two large doors were used for an easy-in, easy-out approach where hay was loaded in the haymow to be thrown down later to the livestock below.

In order to build the barn, the interior siding had to be soaked in the Muddy River before it was nailed in place. A layer of tar paper was added, with lap siding being laid horizontally over that for the final wall. The original cost of the barn was sixteen hundred dollars. In 1919, Fitz added a shed for dairy cows that spanned one-third the distance around the circumference. In 1989, the original barn was reshingled, and the structure was repaired and repainted to its original splendor.

Bruno Barn

TETON COUNTY, CIRCA 1930

Descendants of the first Métis settlers in Teton County built this humble log barn on the South Fork of the Teton River in 1929 or 1930. Mack Bruno Sr. built the barn and a cabin here at that time, with help from Alfred and Clifford Gray. The men harvested the fir logs in Green Gulch and hauled them to the site. The roof boards were milled nearby at Kenny Gleason's sawmill. A typical Métis building, the barn was constructed of rough-hewn logs with saddle-notched corners and a board-and-batten roof.

The Métis people are the children of French-Canadian trappers and traders who married Native women in the late 1800s. The first bands of Métis came into Montana from the Red River drainage and from North Dakota in the late 1870s and early 1880s. Many were seeking a permanent home territory and fleeing persecution by the Canadian government. Having lost the battle for land in Canada, small bands fanned out into Montana, settling near what would later become Lewistown, Cascade, and Choteau.

The Bruno (formerly Bruneau) family and others settled in the canyon near the South Fork Teton River in the 1880s, and the small community thrived for a number of years. About a hundred people lived in the area in roughly twenty-five small log cabins. Most families did not own the land because they were not considered U.S. citizens, but Joseph Bruno (Mack's father) was one of the few who eventually gained a patent for his land, allowing the property to stay in the family. Today, descendants of the original Métis settlers still live in the area, and the rustic log buildings that remain are important reminders of the early inhabitants along the South Fork of the Teton River.

MacDonald/Rossmiller Round Barn

CHOUTEAU COUNTY, 1917

Travelers following the Bootlegger Trail in Chouteau County still stop to marvel at the old MacDonald Barn east of Brady in the community originally named Genou, or The Knees. M. J. MacDonald homesteaded 160 acres here in 1910, and in 1917, after some successful years, he built a round barn. M. J.'s son, Ross, who was eighteen or nineteen at the time, remembered that two railcar-loads of lumber were shipped from Eureka to Brady. A builder named McManus came from Washington to act as the head carpenter.

One of the largest of its kind in the state, this round barn was, in its day, impressive in both size and workmanship. It was framed with two-by-six studs and sheathed with half-inch rough lumber that gently curved to enclose the framing. Four-inch pine flooring was nailed vertically for siding, and small, round "porthole" windows on the second story shed light on the center stalls. The sides of the barn housed cows, and twenty horse stalls stood in the center.

Round barns were something of a phenomenon in the early 1900s, promoted by agricultural schools and Extension Services for their efficiency and sturdy design. The reality was that most Montana farmers could not afford them because of the extra cost of labor and materials. Ross MacDonald said the family built the barn because they thought it would be resistant to the harsh prairie winds in the region. "Otherwise it was not as good [as a rectangular barn]. It had less room for hay in the mow." The MacDonalds used the barn until 1944; it has been in the Rossmiller family ever since. Sadly, the barn has not been used for many years and has deteriorated badly.

Jacobi Barns

CHOUTEAU COUNTY,
CIRCA 1895

The two Jacobi barns have a total of eight zinc ventilators, all built by Harry Jacobi in the late 1890s.

Harry Jacobi was an enterprising rancher who had a diversified operation on the Teton River in Chouteau County. He homesteaded this land along the river in the mid-1890s, just four miles from Fort Benton, raising black silver foxes, cattle, and grain crops. He built the simple gable-roof barns on the property, including the many distinctive zinc ventilators that crown each barn. Originally, the larger barn was sided with vertical rough-milled boards and had rows of windows on the first and second level to provide interior lighting. Over the years, this barn was used for horses, sheep, and milk cows; it has a blacksmith shop at one end. The smaller barn was likely used for foxes and the other fur-bearing animals that Jacobi raised.

Pearson/Tester Barn

CHOUTEAU COUNTY, 1918

The first years of farming for Fred and Christine Pearson near Big Sandy were dry and the work difficult and slow. The tough, compacted sod had to be broken by a one-bottom plow on which the driver rode. Once planted, the crops were slow to grow and the harvests meager. In 1911, the grain was so short they harvested it with a mower. Finally, in 1914 and 1915, the crop was more robust, and in 1916 a bumper crop assured the Pearsons of their success as farmers. That year, they drilled a deep well and built a spacious red barn to house the teams, the milk cows, and the chickens. During the summer, the empty loft was used for dances. The Pearsons' oldest son, Lloyd, spent many pre-breakfast hours in the barn feeding, currying, and harnessing the horses.

Lloyd was nearly killed in 1918 when a tornado destroyed all of the Pearsons' buildings, sparing only their house. Fred and Lloyd saw the storm coming and came in from the field. They put the eight-horse team in the barn. When the storm hit, the high winds raised the barn from its foundation, and the roof landed in the corral beside the barn, trapping a stallion in one corner of the corral. When Lloyd ran out of the barn, the force of the wind blew him down. At the same time, the windmill toppled over, pinning Lloyd under its wheel. He suffered a concussion that left him unconscious for about a month. Luckily, Lloyd recovered, and the Pearsons rebuilt the barn. Today, it is still in use by third-generation descendants of Fred and Christine Pearson: U.S. Senator Jon Tester and his wife, Sharla Bitz Tester.

J. C. Adams Stone Barn

CASCADE COUNTY, 1882–1885

The J. C. Adams Stone Barn is one of Montana's most-beloved and best-known landmark barns. Like the Belmont Park Barn, the Doncaster Round Barn, and Daly's Tammany's Castle (see pages 100, 98, 58), the Adams Stone Barn shows the influence of Kentucky in its design and holds the story of early-day stock raising in its grand walls.

J. C. Adams was born in 1846 in Morgan County, Kentucky, just west of Lexington, the horse capital of the world and home to some of the grandest horse barns ever built. He spent his childhood there working on the family farm before striking out for Montana in 1863. From 1863 to 1874, he hauled supplies to mining camps in Montana Territory. He settled near the Sun River in December 1874, where he began raising sheep, horses, and cattle to sell to cavalrymen at the newly established Fort Shaw.

Adams worked to expand the Sun River ranch, which eventually encompassed three thousand acres; he also opened mercantiles in Sun River and Augusta. By 1882, he was a wealthy

man, able to hire two Swedish stonemasons to build a solid stone barn at a cost of ten thousand dollars—an enormous sum for an outbuilding at the time. The architect of the barn is unknown, but the design, with its central section and matching side wings, towering central ventilator and weathervane, and gracious arched doorways is clearly reminiscent of some of the most luxurious Kentucky horse barns.

Construction of the barn took nearly three years. The sandstone was hauled several miles from a local quarry and hand-cut on-site. To save time and expense, the side walls are rubble stone, and the rear barn wall is wood. The barn was divided into three sections, all entered through beautiful Romanesque arches. The west end held horses and tack, the central section was an open cattle shelter, and the east end was used for storing wagons and buggies but also held a large meat locker. Upstairs, the hayloft floor was finished with long strips of hardwood flooring. Not only was the loft used for storing hay, but it was also used for dances, and occasionally to house travelers along the nearby stage road. To cap off this ostentatious barn, Adams installed a life-sized cutout of a stallion for the weathervane.

Upon the barn's completion in 1885, the *Sun River Press* speculated that it "looks as if it would stand the storms and decay of a century at least," calling it the "Marvel of Montana." Now more than a century has passed. Though the barn has fallen into disrepair several times, each time its owners, the nonprofit J. C. Adams Stone Barn Company, and the Sun River Valley community have rallied to restore the barn in an effort to preserve this "marvel" of architecture and agriculture for another hundred years.

Nevala Horse Barn

CASCADE COUNTY, 1913

(left) Homesteaders in the Korpivaara Finnish community in Cascade County were expert at building with logs.

(below) The logs of the Nevala Horse Barn end walls continue up through the gables—a building design not often carried out because it is easier to side the gable end with milled boards.

The Nevala Horse Barn, built in 1913 by Finnish carpenter Matt Maki, displays a traditional Finnish method of log-barn construction. The two-story barn is made of square-hewn logs connected at the corner with the intricate tooth notch, one of the most difficult corner-notching techniques to achieve. The logs on the first floor are tightly fit; they continue to the roof on the gable ends, where the logs are more rounded, allowing space between them to ventilate the large haymow. The mow floor joists are let through the exterior walls—another hallmark of Finnish log work—and the reinforcing dowel pins are visible in the haymow. The barn was moved from the Nevala homestead to the Lewis homestead in 1945. Roman numerals visible on the northeast wall commemorate that event.

In 1902, Finnish immigrant Gustaf Heikkila was the first to settle along the Little Belt Creek coulee in the foothills of the Highwood Mountains. By 1905, other Finnish settlers had homesteaded the area, calling it Korpivaara, or "potentially dangerous wilderness," for the remote wooded surroundings. There were several builders within the community, including Maki, who brought centuries-old Finnish building tools and techniques with them to the West. Everything from the log harvesting and site planning to the shaping, placement, and tools used to cut the logs followed traditions passed down from generations of previous builders. Having come from a cold and snowy country, Finnish builders in Montana were well versed in building airtight, energy-efficient buildings. Houses, horse barns, saunas, and granaries always had neatly peeled, squared, and tightly stacked logs, with square ends and intricate corner notches. The tighter the logs fit together, the warmer the building in the winter.

Maki used specialized hand tools, including axes, adzes, squares, augers, drawknives, chisels, sheath knives, ripsaws, mallets, and plumb levels. The most important was the metal scribe, or *vara*, which was key in shaping each log to fit tightly. Hand-forged of metal, the *vara* was about ten inches long and featured prongs on one end that could be adjusted to various widths. It was used to trace the natural curve and contours of the upper side of a log onto the underside of the log to be placed immediately above it. The underside of the upper log could then be hewn with an axe following the line inscribed by the *vara*. Finally, a longitudinal groove was cut the length of each log and filled with moss or rags for insulation. This method dates back at least to the thirteenth century and was so effective that it was still in use centuries later when the Korpivaara builders used it on nearly all the Finnish homestead buildings in the settlement.

Heikkila-Mattila Barn

CASCADE COUNTY, 1938

The horse and cattle barn on the Heikkila-Mattila homestead is an intriguing late Finnish-style log barn with American barn design elements. Constructed in 1938 by John Mattila and his brothers, Matt, Richard, and William, the log barn displays Fenno-Scandian log-wall construction with skillful toothed-notched corners. This type of notch was the most intricate of Finnish log-wall construction techniques and also provided for a very tight wall, eliminating the need for chinking.

While the log walls display much skill and tradition, the Heikkila-Mattila Barn design departs from ethnic traditions in its use of rounded logs and a gambrel roof. Most Finnish barns were typically one-story with a hayloft under the gable roof. The Mattila brothers were second-generation Finns who had been born in Montana and who traveled the area from Belt to Butte for work. In their travels, they must have observed the large, gambrel-roof barns that were popular in the early twentieth century. They may also have been influenced by farm design bulletins issued by Extension agents who made regular visits to Belt.

The beautiful and masterful log work on the Heikkila-Mattila Barn was accomplished using only handsaws and chisels.

Jemison Barn

CASCADE COUNTY, 1918

David D. Jemison bought property in the Chestnut Valley in 1918 and purchased plans for a house and barn from the Sears, Roebuck and Company. The materials were all precut to specifications and delivered to the farm site with complete building instructions. With the help of relatives and neighbors, Jemison built this spacious livestock barn and a comfortable Craftsman-style bungalow.

Innovations in barn design and the ease with which goods could be transported by rail in the late 1910s allowed farmers to build barns that were not rooted in an ethnic or family building tradition. Pattern-book barns such as the Jemison Barn were the first "kit barns."

Sears, Roebuck was the leader in mail-order housing; it also offered several barn designs in its home catalogs. Beginning in 1918, Sears offered *The Book of Barns*, a catalog exclusively dedicated to barns. The company continued to offer barn plans through 1929. Other midwestern companies, such as Montgomery Ward, James Manufacturing Company, Alladin, and Gordon Van Tine, offered barn kits from the early 1900s generally up until World War II.

The ranch owners Joe and Leslie Tripp plan to preserve the structure so future generations can enjoy it as well.

Current resident Tricia Gibbon-Siler raises sheep on the ranch and says, "I can't imagine life without a barn."

Ayrshire Dairy Barn

CASCADE COUNTY, 1912, 1914

(left) Ayrshire Dairy, 1940. In the foreground are (left to right) Harry Mitchell (age eight), his father, Fergus, and Fergus Jr. (age fourteen).

(opposite) The original dairy barn, built in 1912 by Harry Mitchell Sr., was enlarged in 1914 and became the horse barn.

When Scottish immigrant Harry B. Mitchell bought land south of Great Falls in 1906, he brought in Ayrshire Highland cows to start a dairy that became one of the largest in north-central Montana. With help from his son, Fergus, and daughter-in-law, Harriet, Mitchell developed the Ayrshire Dairy into a state-of-the art operation, handling every aspect of milk production, including processing and distribution. The north end of the three-gabled barn is the original dairy barn built in 1912; it was enlarged and became a horse barn when the long, monitor-roof dairy barn was built in 1914. Horses and wagons were used until 1937 to deliver milk in town, after which a fleet of milk trucks ran regular neighborhood routes.

The Mitchells designed the spacious 1914 diary barn for a large herd of dairy cattle and for maximum light, ventilation, and sanitation. Its first-floor windows provided more light and air circulation to the cows than a standard gambrel-roof barn, and concrete floors and half-walls in the stanchions allowed waste to be easily washed into gutters and down into sewer drains. *Radford's Practical Barn Plans* of 1909 marketed a "Model Cow Barn . . . Designed very carefully to provide every comfort to a herd of thoroughbred Guernseys." The Ayrshire Dairy Barn is quite similar to the Radford "Model Cow Barn," right down to the twin silos at the far end.

The Ayrshire Dairy was among Montana's longest-lived dairies. It was one of the first dairies in the state to pasteurize its milk and was the first to homogenize milk. Great Falls had about twenty dairies through the teens and 1920s; only five survived the Great Depression through to the 1950s. Fergus's son, Harry B. Mitchell II, took over the business in the late 1960s and for many years competed successfully with local dairy giant Meadowgold. Eventually, consolidation in the dairy business forced the closure of the Ayrshire Dairy, and in 1990, after eighty years, the Ayrshire cows were sold. Today, the 1914 barn still stands, awaiting the future.

Ayrshire built a new dairy barn (left) in 1914, with rows of windows on both floors to provide maximum light and ventilation.

Albright Ranch Barns

CASCADE COUNTY, 1940

Nestled into Deep Canyon, the barns on the Albright Ranch represent two completely different eras of livestock ranching. The long, weathered wood barn with the low-pitched roof was a sheep barn on the original ranch, established by William Albright in the late 1890s. Albright ran several bands of sheep here, and the ranch became the town site of Albright on the Great Northern Railway. In its day, the town consisted of a few houses, a hotel, and a boardinghouse. In addition to his sheep operation, Albright also had some mining claims, a limestone quarry, and a kiln.

Albright sold the ranch to Charlie and Sue Bovey in 1937. The couple is well known in Montana for preserving the mining town of Virginia City and re-creating Nevada City by moving in many log buildings that otherwise would have been burned or torn down. In 1940, the Boveys had two show barns constructed on the ranch to look like pioneer-era livery barns. These two identical barns enhanced the pioneer town atmosphere at the ranch and most likely were influenced by the large, false-front livery barn not far away at the old town of Heath (see page 183).

(above) The false-front barns on the Albright Ranch were sale barns, deliberately built to look old for well-known Montana history buffs Charlie and Sue Bovey.

(right) Despite its current equine occupants, this long barn with its low-pitched roof originally served as a sheep shed.

S. S. Hobson Ranch Barn

JUDITH BASIN COUNTY, 1890s

This nineteenth-century barn near Hobson is associated with prominent rancher and Montana senator Simeon S. Hobson. Hobson bought the ranch in 1892 and likely supervised building of the house, the central section of the barn, and several outbuildings throughout the mid-1890s. The barn is a good example of a center-aisle barn with a wide door under the gable end and horse stalls lining each side of the alley. In 1915, Hobson retired and sold the ranch to Peter and Sylvia Campbell and sons Harold and John, who built side and rear additions. The silo was built in 1926 to augment the dairy operation.

The majority of Montana barns built in the late 1800s were utilitarian log or stone structures with few windows and ventilators; they were usually devoid of any ornamentation.

The S. S. Hobson Barn stands out for its whimsical decorative details, distinctive cupola, fine craftsmanship, and finishes such as double-hung windows and clapboard siding, which were usually reserved for residential construction. Indeed, the buildings on the ranch reflect Simeon Hobson's status as a businessman, gentleman rancher, and Montana legislator.

Hobson came from Maine to the Judith Basin in 1880 and began ranching in 1882. He eventually amassed over five thousand acres in the Judith River Valley. He was an organizer of the Fergus County Bank in Lewistown, an active member and founder of the Montana Stockgrowers Association, and a Montana state senator and delegate to Montana's constitutional convention in 1889. In 1907, Hobson donated land for the Hobson Railroad Station on the new Billings & Northern Railroad, a branch line of the Great Northern. The new station quickly supplanted nearby Philbrook, and after some debate, facilities were moved to the new town, which was given the name Hobson. The ranch is now owned by the John Campbell family.

Meadowbrook Stock Farm Barn

JUDITH BASIN COUNTY, 1917

Scottish-born homesteader Thomas Murray came to the Judith Basin in 1883 and settled on 160 acres, running cattle and sheep on the open prairie. By the turn of the century, Murray's ranch, known in its heyday as the Meadowbrook Stock Farm, was given to wheat farming and to the breeding of cattle, hogs, sheep, and horses. The coming of the railroads in 1906 and 1908 promised better transportation to agricultural markets and sparked new towns such as Hobson and a desire for more substantial buildings.

This imposing, gambrel-roof barn, along with a sheep barn and other outbuildings, was constructed in 1917. The barn is arranged around a central alley with four double stalls on each side and a feed trough near the center. A tack room, feed room, and large box stall occupy the rear. The cavernous, two-story loft was equipped with four tall hay chutes for dropping bales to the barn below.

Although Murray left the ranch in 1924—a victim of drought and horrendous pest infestation—the ranch today retains its original appearance, an excellent example of an early Montana farming enterprise.

The barn at Meadowbrook Stock Farm had badly deteriorated by the time owners Jean and Michael Trammel restored it in 2002.

Central Montana Agricultural Experiment Station Barn

JUDITH BASIN COUNTY, 1908, 1921

Although converted for use as a research laboratory, the original 1921 dairy barn has seen few changes on the exterior.

The barns on the Central Montana Agricultural Experiment Station were built during the heyday of central Montana's dryland homestead rush. The station was part of an agricultural experiment station system for farmers created by the 1907 Montana legislature. One of seven research stations built in the state, the center played a prominent role during the homesteaders' bonanza years, then evolved along with the gradual maturation of dryland farming into a staple component of Montana's agricultural economy following World War II.

Over the years, the buildings have been adapted to fit the needs of agricultural research and technology. The original horse barn (not pictured) was built in 1908 and used to house draft and stud horses. Later, it served as a garage; currently, it is a seed-processing lab. The dairy barn, built in 1921, employed the latest milking technologies, but in the 1950s it became a machine shop. Then, in the mid-1990s, it was converted to a dry lab and meeting spaces. Most recently, the milking parlor was converted into a wet chemistry lab, where cutting-edge ethanol research is now under way.

Heath Livery Barn

FERGUS COUNTY, 1908

The town of Heath sprang up in 1908 when Perry Heath established a stopping place in the foothills of the Snowy Mountains to serve those traveling to and from Billings. Heath opened a store, a post office, a schoolhouse, and a saloon. He also hired Sam Bowman and John Plank to build a false-front livery barn with room for more than a dozen horses and with a large sliding door to allow carriages and wagons to pull right in. The loft featured a red fir wood floor, and for years, old-timers in the area fondly remembered Saturday dances in the barn's loft.

This oasis in the foothills of the Snowies was a welcome stop for teams and drivers navigating through the mountains. Frequently, riders would leave their horses in the barn and catch a train ride into Lewistown after the Milwaukee Railroad came through the town in 1913. The U.S. Cavalry and Stagecoach line coming from Billings by way of Grass Range regularly stopped at the barn to take advantage of the refreshing natural spring water. In 1917, much of the town of Heath was devastated by fire, but the barn was spared and remains in good shape today. Since 1935, the barn has been used for ranch operations.

The Naylor Big Barn

FERGUS COUNTY, 1926–1927

The Big Barn on the Naylor Brothers Ranch was built by Axel Johnson, a proud craftsman who built many barns, houses, and buildings in the Harlowton area. Johnson built the barn for Floyd Naylor, a farmer who raised sheep, cattle, and hogs for market and was an active participant in a dizzying array of Extension Service programs designed to improve farm and ranch production. In the mid-1920s, Naylor raised demonstration crops, participated in cattle and sheep finishing demonstrations, sat on Montana State College's (today, Montana State University's) Extension program livestock committee, planted shelterbelts, and hosted field days at the ranch to showcase the newest farming information and techniques.

Naylor was featured in Fergus County Extension Office reports as an advocate for early lambing. His sheep work is also the subject of another document, the "Fergus County Agricultural Development Program 1928," which contains a

The Naylor hog barn reflects a typical Extension Service design.

photograph of the 1926–1927 construction of the Naylor Big Barn. The photo's caption described the barn as "a modern lambing shed under construction by Floyd Naylor of Buffalo. Mr. Naylor has tried to make it most convenient, thus reducing costs and increasing livability of lambing."

Naylor was involved with agricultural extension programs right up through 1931, when the ranch was foreclosed upon. Years later, Axel Johnson returned with his son to visit the Naylor place. After a first glance, Axel grumbled, "Those damn guys said they would put concrete under the barn in the summer." Because it had been winter and too cold for concrete when Johnson had begun to build the barn, he had put it on wood pilings with the Naylors' agreement that they would put in a concrete foundation the next summer. Though that never happened, the barn has stood proudly on wooden pilings for eighty-plus years.

The Naylor Brothers Ranch was listed on the National Register of Historic Places on December 20, 2008. Owner Corey Merrill received a grant from the Montana State Historic Preservation Office in 2009 to reroof the barn. Concrete footers were put in place at the same time.

Smith Barn

FERGUS COUNTY, 1882

Christian Barr Smith came to the Beaver Creek Valley in Fergus County in 1882, following many years in the Helena Valley raising beef and dairy cattle for miners. Claiming 160 acres under the 1862 Homestead Act, he began making improvements on the land. He built the center section of the barn from six-by-six fir timbers fastened with wooden pegs in skillfully crafted mortise-and-tenon joints. A lean-to storage area on the west side and a milking room on the south were added later. Five generations of Smiths stabled their horses, stored hay, and milked dairy cattle in this barn. In the early years, the Smiths raised draft horses and Hereford cattle; today, they run Black Angus–Simmental cattle here.

This humble-looking wooden barn is a fine heavy-timber frame barn; its weathered gray wood, stout timbers, and rusted nails represent the hopes and the hard work of one family, rooted in this place for more than 120 years.

Peeled saplings are the battens between the boards on the handsome façade.

Lingshire Barn

MEAGHER COUNTY, MID-1910s

This immaculate log barn is an excellent example of fine carpentry and the creative use of local materials. It was put up by Henry Lingshire and a crew of builders in the mid-1910s for the Walker Land & Livestock Company located northwest of White Sulphur Springs. The logs were cut near the ranch from what they called the "pole patch" up Rock Creek, where lodgepole pines blanket parts of the drainage.

Working with the relatively thin and short pines from the "pole patch," Lingshire and his crew combined different joinery and siding techniques to give this barn its unique character. On the first floor, Lingshire's crew squared the logs on both

Details of the vertical channel posts and stacked logs on the sides of the barn

sides, then notched and stacked them into vertical channel posts. This *piece-sur-piece* construction, which descended from territorial-era trading-post builders, allowed a large structure to be built from short lengths of logs. For this barn, instead of using channel posts at the rear corners, the builders square-notched the ends. On the façade, small-dimension logs and saplings were used to create rustic board-and-batten siding. The rear gable and upper sides of the barn were then sided with machine-sawn board-and-batten siding.

Inside the barn, meaty two-inch-thick floorboards, heavily grained posts, and the ethereal glow of the afternoon sun give a sculptural quality to the stalls. It may be clichéd to say they don't build barns like they used to—but this barn proves that it's true.

Charles M. Bair Ranch Barns

MEAGHER COUNTY, LATE 1920s

The Bair Family Ranch, once one of Montana's largest sheep ranches, is today one of the state's most authentic museums, known for the trove of international artworks inside the house and its documentation of the family's prominent history in ranching culture. The Bairs' horse and sheep barns reflect the evolution of the ranch and the family's important role in the business end of the sheep operation.

The old horse barn is now the Bair Ranch Museum visitor's center.

One of the old sheepherder's wagons is parked at the ranch.

The Bair Ranch still raises about three thousand sheep. Activity in the barn is a blur at shearing time.

In 1890, Charles M. Bair bought his first ranch near Lavina, but he sold it in 1893 and moved his family to Billings. He continued to ranch on leased lands near Billings and in 1898 ventured for a time into gold mining in the Klondike. Bair soon returned to Montana with enough money to expand his sheep operation. At its peak in the early 1900s, the Bair Ranch ran between two hundred thousand and three hundred thousand sheep on land that stretched from the Crow Reservation north to Forsyth and west to Martinsdale. In 1910, Bair lost his lease on the Crow Reservation and began to focus operations on his Musselshell River lands. In 1913, he bought John Grant's sheep ranch near Martinsdale and expanded the ranch. He owned about sixty thousand deeded acres and leased about the same amount. The Bair Ranch was reportedly Montana's largest sheep ranch until the Great Depression forced Bair to reduce his herd.

Horse barn at the Bair Ranch headquarters

The huge sheep barn has always been used for lambing, inoculating, and shearing the ranch's sheep. Bair built the big white gambrel-roofed barn near the main ranch house in the 1920s to shelter workhorses; the ranch later used it for wool storage. In 1993, it was remodeled and converted to a visitor's center for the museum. Another complex of white barns, built in the 1920s, is located about a mile north of the ranch house at what is now the ranch headquarters.

Sims-Garfield Ranch Barns

GOLDEN VALLEY COUNTY, CIRCA 1886

The log and stone buildings on the Sims-Garfield Ranch near Ryegate are some of the oldest on the Musselshell River. Originally settled in 1882 by John T. Sally, the ranch was taken over by Joseph and Mary Sims in 1885. Under the pioneering ownership of the Simses, the ranch grew to include a six-room log house, an impressive entrance gate, and two barns: a large log barn and a stone barn.

The Sims-Garfield Ranch stands below a long line of rimrocks marking the shore of an ancient lake, and the soft stone that litters the area was used in many residences and outbuildings. It appears that Sims began constructing the barns in the mid-1880s and continued over the course of years until the two sections finally met in the middle. The log barn is a fine multicrib barn, built by connecting six one-room sections

Owners Jess and Nicky Garfield

under one roof. It has horse stalls at each end and an open area in the center. The end of the log barn shares a wall with the stone barn, but there is no door connecting the two buildings. The stone barn includes a bull pen at one end; the other half is open for cattle loafing or storage.

At the time the Simses purchased the ranch in 1885, the Musselshell Valley was open livestock range. Managing to weather the devastating blizzards of 1886–1887, the Simses rebounded, and by 1900 the ranch employed at least ten hands. In 1906, with talk of the coming of the railroad, the ranch was sold to land speculators, who in turn sold a one-hundred-foot corridor of ground to the Chicago, Milwaukee & St. Paul Railroad. The new railroad was completed in 1908, and the siding located here was named Ryegate after the "fine field of rye" planted by Sims and admired by railroad officials.

The ranch was leased to Harry Henton and Verne Johnson in 1923. Henton eventually became sole owner. Later, his daughter and grandson, Emmy Lou and Jess Garfield, took over. Today, the forty-one-thousand-acre ranch is listed in the National Register of Historic Places.

Lindstrand Barn

MUSSELSHELL COUNTY, 1908

When Magnus Lindstrand immigrated from Sweden in 1886, he settled west of Roundup and began ranching the following year in what was then a sparsely populated place, accessible only by horse and wagon. Two decades later, the Milwaukee Railroad surveyed for its new line right through the Lindstrand place, so the family cabin was abandoned, and a new house and barn were constructed away from the tracks. Built in 1908, the barn held horses, sheep, and cattle, with space for hay storage in the loft. The house and barn have been landmarks along the Musselshell ever since.

In 1951, the *Roundup Record Tribune* offered a poetic obituary to Magnus Lindstrand, perfectly describing the spirit of many Montana ranchers who made their living from the land and left a legacy for the next generation to build on. "Mr. Lindstrand," the notice ran, "was of that type of stalwart immigrant whose determination to settle here was a boon of inestimable value to America and who contributed vastly to the greatness of our country. He reared a family of fine men and women whose progeny in turn bid fare to enrich the land with upstanding citizenship. He faced adversity and hardships without flinching, and with persistence and earnest, honest effort built a home and way of life that he hoped might endure."

Thirty Mile Stage Barn

MUSSELSHELL COUNTY, CIRCA 1890

Andrew Shannon, an Irish rancher, settled this Musselshell Valley ranch before 1885 and opened a livery barn here to serve the stage line operating between Billings and the Musselshell. In 1896, Shannon hired James H. Buckey to manage the property; a roadhouse, saloon, and post office soon stood next to the livery stable.

The barn is an exceptional example of a large *piece-sur-piece* log barn. Typically associated with trading posts and French-Canadian builders, this method of log construction used weight-bearing corner- and medial-wall posts into which horizontal wall logs were carefully fitted and secured with tenon joints. This method produced an extremely tight wall and allowed the builder to use shorter lengths of logs to build a barn of any size. The Thirty Mile Stage Barn is two sections wide and three sections long, much larger than most barns of this type.

For a short time, the stage stop was known as Buckey, Montana, but its location thirty miles from Billings ultimately gave it the name that stuck. While the railroad was being built through the Musselshell, Buckey and his family kept busy serving forty to fifty men. Completion of the Chicago, Milwaukee & St. Paul Railroad line south of Roundup put an end to the stage line in 1908, although Buckey continued to deliver mail until 1915. The Thirty Mile Ranch kept this fine barn in use until the late 1980s. Before it fell out of service, it had housed horses and ranching activities for a century.

Bussell Farms Barn

TOOLE COUNTY, 1919

Doug and Ruby Hellinger Bussell's family established their thirty-eight-hundred-acre Montana farm in 1913. The first substantial building that went up on the property was a state-of-the-art round-arch barn, erected in 1919. Doug and Ruby say that "although Grandma Hellinger would have rather had a home to replace the one-room shack that housed nine people, the family knew that a proper barn was critical to the survival of the livestock and successful farming."

The Bussells used the barn for workhorses, calving, tack and feed storage, and in the loft, hay storage. The wide drive-through alley allowed them to drive a team in one door, load manure or unload feed, and continue on out the back door. The large braced ladder in the hayloft provided a safe way to climb to the top of the hay bales to throw a bale down a chute to the horses below.

The family has worked to keep the barn in good condition over the decades. Originally left untreated, the barn's wood siding received an initial coat of paint in 1965, necessitated by the harsh Montana winters. Since then, the family has replaced the roof, repaired interior joists, and repainted three times. The barn is still the centerpiece of the farm, used for saddle horses, calving, small tractor storage, and hay storage. The Bussells received a National Trust for Historic Preservation Barn Again! Recognition Award in 2007 for their efforts.

Fort Assinniboine Stable 4 and Bull Barn

HILL COUNTY, 1906

Fort Assinniboine was built in the aftermath of General G. A. Custer's demise at the Little Bighorn. It was intended to ward off possible attacks by Indians who had escaped to Canada at the end of the Indian wars. At the time of its construction, there was concern that Sitting Bull's people would return from the Cypress Hills or that the Nez Perce, following the defeat and capture of Chief Joseph in the Bears Paw Mountains in 1877, would retaliate against white communities.

In its day, Fort Assinniboine was the most elaborate post in the United States. It boasted 104 staunch brick buildings constructed over a four-year period. The bricks were manufactured on-site under the direction of Colonel C. S. Broadwater, and much of the building was done by the troops themselves. Construction progressed so rapidly that the Cheyennes said that Fort Assinniboine "rose out of the ground."

The long brick building is the last of six cavalry stables at the fort; it was built in 1906 to replace the original stable, which was built in 1877 and destroyed by fire. The stable is huge—233 feet long and built to house approximately seventy-three horses. Nearby there were three duplex cavalry guardhouses and blacksmith shops, each with quarters for two blacksmiths and two identical shops at the rear. Sometime after Fort Assinniboine closed in 1911, the stalls were removed, and half of the stable was remodeled as a community events center.

The other massive "barn" at Fort Assinniboine is now known as the Bull Barn by the Northern Agricultural Research Center. Built in 1908, it was not intended to house animals or feed but was built as a thousand-ton icehouse. The foundation is stone, and the upper walls are several withes thick to provide insulation. The columns at intervals along the exterior wall are buttresses to reinforce the walls and hold the weight of the ice.

The Tenth Cavalry, composed of African American soldiers, dubbed "buffalo soldiers" by the Cheyennes, arrived at the fort in 1892. The black troops of the Twenty-fourth and Twenty-fifth Infantries were also stationed here and were likely the most frequent users of this stable and bull barn.

(opposite) 1906 cavalry stable

(below left) The ice house, which is now known as the Bull Barn

(below center) Inside the former ice house

(below right) This area of the cavalry stable was converted for use as a community center.

Burns Ranch Barn

BLAINE COUNTY, 1910

From the traditional red siding and gambrel roof to the white trim, matching ventilators, and tall metal windmill, the barn on the Burns Ranch is the embodiment of the classic American barn.

Homesteader John Archer bought this property in 1901 and built the barn and several outbuildings in 1910. It was well designed, with stalls for horses and cows lining the central alley at the front; an open area for storage or calving in the rear; and a steel track, which is fastened to the rafters and runs through the center, to carry manure buckets out the barn's back

door to be hauled away. Archer survived on his homestead for over a quarter century but gave up the struggle in 1937, selling the farm to the U.S. government under a Depression-era resettlement program.

Sam Burns bought the property from the federal government following World War II and began a cattle and wheat farm. New methods of cattle management and soil conservation helped to rebuild and expand the operation, which is still in the Burns family today. It stands as a reflection of the optimism of new farmers coming to eastern Montana in the early 1900s and is a landmark much photographed by travelers along U.S. Highway 2.

Joe Hartman Barns

PHILLIPS COUNTY, 1914

In the rolling Camp Creek Valley southeast of Malta, the barns on the Joe Hartman Ranch tell the story of mining, ranching, and freighting across the prairies and the rugged Little Rocky Mountains in the early 1900s. Joe Hartman holds a special place in the history of Zortman, for he was a legendary freighter who was a lifeline for folks living in that small Little Rockies mining town. From 1904 to 1916, Hartman regularly hauled provisions from Malta to miners in Zortman, a 110-mile round trip difficult even under the best of conditions and treacherous when the weather turned bad. With as many as sixteen draft horses in his team, he pulled four wagons loaded with up to fourteen tons of food and mining supplies. He also carried ore from mines high in the mountains down steep and dangerous grades to the mill in Zortman.

Joe Hartman is remembered for saving Zortman from a bleak Christmas in 1906, when the townspeople had little food and drink and a dwindling supply of mining equipment. Having left for Malta in late November that year, Hartman had just begun his return trip when the first winter blizzard hit. Undeterred, he continued on through deep snowdrifts and minus-sixty-degree windchills, arriving in Zortman on Christmas Eve, forty-four days after he had left.

When Hartman wasn't freighting, he was at his ranch southeast of Zortman, allowing his teams to rest. In 1914, he designed and built an unusually large horse barn to accommodate his twelve-hundred-pound horses. The stout barn was built with oversized eight-inch-by-ten-inch timber support posts cut from nearby Coburn Butte. The hayloft could hold

over 125 tons of hay and once hosted more than 250 people for a barn dance.

After retiring from freighting in 1916, Hartman began raising draft horses and sheep. In 1928, he built a gigantic sheep shed for his band of twenty-five hundred sheep, using plans provided by the U.S. Sheep Experiment Station in Dubois, Idaho. The barn, described at the time as the "most modern sheep shed in the state," measures 80 feet wide by 175 feet long and has a three-tiered roof with numerous south-facing windows to allow for solar heat gain in the winter. Hartman's shed and his sheep were also featured in a 1929 issue of the *Montana Farmer-Stockman* about modern sheep ranching. When he died in 1971, he willed his ranch to the state for use as an agricultural experiment station, extending his legacy as a rancher in perpetuity.

(left) Draft horse barn

(above) Hartman's fully modern 1928 sheep barn

Jefferson Dredge Barn

VALLEY COUNTY, 1941

The Norman children with the family dog, Calley (left to right): (on bikes) Mary Emma, Joshua, and Hannah; (standing) Brett, Grant, and Mitchell

The blue-and-white dredge walls were hauled to the Reddig place in one piece.

The roof sheathing reveals how the barn was finished with boards from many different parts of the dredge.

One of the most fascinating accounts of a farmer's ingenuity is the story of the Reddig barn and how one ambitious farmer, his sons, and the community helped move the Jefferson Dredge thirty miles downriver and then north by truck, where it was reconstructed into the Reddig family barn.

The Jefferson Dredge was one of four barges used to build the Fort Peck Dam across the Missouri River between 1933 and 1940. The dam, one of the largest hydraulically filled earthen dams in the world, contains 125.6 million cubic yards of fill, most of which was sucked up from the floor of the river by giant augers attached to the dredges. In 1939, with the dam nearly complete, three of the four dredges were auctioned off, and Karl Reddig was the highest bidder for the Jefferson Dredge, paying $251—or just over $3,800 in 2009 currency.

During the winter of 1940, while the dredge was still frozen into its parking spot, the Reddigs dismantled its walls and hauled them to their farm near Lustre. Then, Karl Reddig perched himself atop a tractor seat attached to one of the dredge supports and, with the help of his sons, Wilmer and Calvin, piloted the dredge, negotiating many bends and curves, thirty miles downriver. From a point south of Frazer, the Reddigs dismantled everything but the hull and hauled the salvaged materials to the farm. The hull remained in place (and is still visible) because it was so sturdily built that it was impossible to break up and haul away. Karl, having already prepared a design, then hired a carpenter to help complete the barn.

The barn was finished in 1941, near the end of many years of drought and a depressed economy nationwide. Looking at the barn today, one can easily see where whole sections of the dredge walls were used to create side walls for the barn. All of the windows came directly from the dredge, and the rafters in the hayloft still show the dredge's original blue and white paint. Today, this barn stands in tribute to the people who came through the lasting hardships of the Great Depression and is an excellent example of the resourcefulness and frugality of the Montana farmer.

Etchart Stone Barn

VALLEY COUNTY, 1917–1918

Mitch Etchart continues the legacy of his father, John, who began ranching southeast of Glasgow in 1912.

One of the enduring landmarks of rural Valley County, the Etchart Stone Barn stands majestic amid the grassy hills at the head of Willow Creek. Built by John Etchart and local stonemasons in 1917–1918, the barn is a fine French-style stone barn.

Etchart immigrated to the United States from the Basque Pyrenees area of southern France, where sturdy stone houses and barns were the norm. Not only did Etchart want his homestead to reflect the Old Country, but it was also the most logical and economical way to construct his farm buildings. The sandstone was readily available in the nearby hills for the cost of hauling it to the site by horse-drawn stone sleds, a far better option than purchasing lumber and hauling it over fifty arduous miles of primitive trails on horse-drawn wagons. Etchart constructed the barn and ranch house with the help of the Partch brothers from Saco, stonemasons who also helped build the Sleeping Buffalo Resort between Saco and Malta.

Lee Barn

ROOSEVELT COUNTY, 1917

The Lee homestead buildings represent the effort of John Lee, one of many Norwegian immigrants to eastern Montana in the early 1900s who struggled at diversified farming in this semiarid country. The barn sheltered draft horses and a few dairy cows, held tack and other farm equipment, and had space for hay storage in the loft. Inside, the barn has a side-pass plan instead of the wide central alley seen in so many barns of the era. It has stalls back-to-back right through the center of the barn, with wide alleys along the outside walls and a very narrow walkway through the center. This was a common design in many European barns throughout the 1800s.

The remains of this homestead and many others like it are scattered across the region, a reminder of the hundreds of optimistic immigrants to the area who tried to make a go of dryland farming but lost the struggle. Between 1900 and 1910, rain was plentiful, and the number of farms in Dawson County (which included all of present-day Roosevelt, Garfield, McCone, Prairie, Richland, and Wibaux counties) increased from 226 to 1,946. Many settlers were lured by railroad promotions, land speculators, the federal government, and the promise of dryland farming. Early successes during the wet years and the theories advanced by the dry-farming movement's spokesman, Hardy Webster Campbell, helped hide the fact that eastern Montana was a semidesert grassland.

Over the course of twenty years, this country was developed into hundreds of small farms—to disastrous effect. Many immigrants who risked everything to move to Montana ended up leaving more impoverished than when they came. The evidence of many homesteads has since been plowed under, and remaining places like the Lee homestead are important icons of Montana's early homesteading history.

The skinny alley, or side-pass alley, in the Lee Barn was a popular barn design throughout Europe in the 1800s.

Ten-year-olds Richard Kerstein (center) and friend Ralph Fisher posed in the 1950s with the Kerstein's farm dog, Rinty, named after the popular TV dog Rin Tin Tin.

Kerstein Barn

DANIELS COUNTY, 1914–1916

The casual observer of the Kerstein Barn near Scobey might not notice anything unusual about this barn, but a seasoned rancher would say right away, "That barn doesn't have a hay door. How do you get the hay into the loft?" The answer is, "You can't."

The Kerstein family built this gambrel-roof barn as a granary and dance hall between 1914 and 1916. Its sixteen-foot-wide doors and concrete floor made it possible to drive wagons into the barn to unload grain and cargo into the grain bins, and some say that the Kersteins allowed bootleggers to store whiskey brought from Canada (just fifteen miles north) in the easily accessed bins. Upstairs, the loft has never seen hay; instead, its floor is covered with golden tongue-and-groove hardwood, perfect for a two-step or even the Charleston.

Richard Kerstein, here with his grandchildren, hopes to restore his family barn and keep the legacy of his father's generation standing for his descendants to use and enjoy.

Like many city folks lured by the promise of free land, the Kersteins found life in the country to be an adjustment. Emigrating from the factory district of Chicago in 1910, the family took up a homestead near Four Buttes. Current owner Richard Kerstein says his grandfather, father, and uncles were farmers, not cowboys, and the first in the area to pull a plow with anything but horses. As soon as they could, they sold their horses and bought gasoline-powered trucks and tractors.

As for the barn, bumper crops of the early 1910s may have set the family's hopes for the future too high. Or, as Richard Kerstein speculates, perhaps the family wanted a dance hall but couldn't justify the expenditure, so his father, Sidney, an amateur musician, built a granary large enough to hold dances. Whatever the reason, the Kersteins had many great barn dances, but the grain bins were filled for only a few years before drought and the Great Depression sent most of the family back to Chicago. Sidney and his brother Sam persevered, however, and today the barn remains in the Kerstein family.

Hughes Barn

DANIELS COUNTY, 1916

On June 15, 1904, the Frank Hughes family spent their first night as squatters on land they eventually claimed for a homestead. They had come by train from Monticello, Indiana, to Poplar, Montana, and while Frank Sr. scouted the area, the family camped out in a tent in Poplar. When Frank returned, they assembled a Sears, Roebuck wagon, bought horses from an Indian friend, loaded the supplies they had brought with them by emigrant car, and set out across the prairie to the site Frank had chosen.

Frank Hughes was the first farmer in the area to succeed in dryland farming, raising three acres of oats in 1906. His daughters gathered wild flaxseed for several years, and Hughes planted and raised an excellent crop of flax in 1912. Cattle, pigs, horses, turkeys, geese, and chickens were introduced to the venture, and when homesteading opened on the Fort Peck Reservation in 1913, the Hugheses claimed the land they had settled on.

The Great Northern Railway ran a spur line through the area in 1913, making lumber and supplies more readily available. In 1916, Hughes hired a seven-man crew of Swedish and Norwegian carpenters to build a home and barn. Completed in one summer, the barn cost fifteen hundred dollars.

After the Hughes farmhouse and barn were completed in 1916, the Great Northern arranged to photograph the farm. A camera crew came at harvest when the grain shocks were still sitting in the fields. To make the area's farming possibilities enticing, the crew moved numerous shocks into the picture's foreground to make it appear as though the harvest had been more abundant than it actually was. The picture was used on posters by the Great Northern Railway to encourage settlement in the Daniels County area.

CHAPTER 6

Barns of the Yellowstone Basin

MONTANA'S Yellowstone River is the longest free-running waterway in the United States, 670 miles of river that shimmers its way from Yellowstone National Park to eastern Montana. The beaver-rich river and the buffalo-filled plains drew fur traders like Louis-Joseph and François de La Verendrye, the first Euro-Americans known to venture beyond the Mandan-Hidatsa villages on the Missouri River to the Yellowstone in 1742, searching for the elusive route to the "western sea."

In 1805, French Canadian François-Antoine Larocque traversed the heart of the Roche Jaune, or the Yellow Stone, River region, as did William Clark in 1806. It was prospects for a rich fur trade that most impressed them, as Clark observed: "for me to mention or give an estimate of the different Spcies of wild animals on this river particularly Buffalow, Elk, Antelope & Wolves, would be increditable."[1]

Just one year later, fur trader Manual Lisa's crew built Fort Remon at the mouth of the Bighorn River, the first trading post in Montana.[2] Other forts followed. On the Yellowstone, they included the first Fort Benton (built near the site of Fort Remon in 1821) and Fort Union (at the mouth of the Yellowstone in 1828). These were the heady days of the fur trade era, when supplies of beaver pelts and buffalo robes seemed limitless. A string of forts to guard the trade came and went, and by the 1860s the decline of abundant beaver pelts and buffalo hides was imminent. Life and enterprise on the plains were about to change.

The Yellowstone Headwaters

The upper tributaries of the Yellowstone wind their way around mountain ranges and through valleys that bear poetic names—Sweetgrass, Stillwater, Paradise. Carved by the force of the Yellowstone, these are the high-elevation valleys into which miners and ranchers entered during the later 1800s, prospecting for gold and claiming open ground upon which to launch agricultural pursuits.

The Emigrant Gulch mining camp in Paradise Valley took hold in 1864, and that year Ben Strickland planted wheat nearby, joining agriculture to mining. The mining camp boomed and

(opposite) Baum Barn, Yellowstone County, built circa 1912

busted but not before agrarian settlement expanded. Boteler's Ranch was established at the mouth of Emigrant Gulch in 1868. An isolated outpost high up the valley, it quickly became a stage stop for travelers, including the 1871 Hayden Expedition en route to view the oddities and geothermal features of the Yellowstone Plateau.

The plains of the lower Yellowstone Basin were still Indian territory, reserved for hunting and traditional life. However, displacement by war, mining, and white settlement constrained Native traditions. As elsewhere, military posts and Indian policies introduced new agricultural pursuits to the region. In 1869, Fort Parker, the first Crow Indian Agency, was built at the mouth of Mission Creek, just east of present-day Livingston. The fort's agents provided tools and encouraged the Crows to learn to farm. As historian Carroll Van West relates, "Agent Pease had his men construct three miles of fences and two miles of irrigation ditches while planting one hundred acres of good bottomland by the summer of 1871. These actions marked the beginning of the valley's agricultural cultivation."[3]

In 1873, the steamship *Coulson* ventured up the Yellowstone to chart the way for transportation. Two years later, the steamboat *Josephine* chugged its way to a point near modern-day Billings, penetrating deep into the Yellowstone Valley, and that same year the first permanent settlement along the lower Yellowstone was established at Countryman's Landing (now Columbus, Montana). Military camps at the Bighorn and Tongue rivers and campaigns against the Sioux and Cheyenne secured the region for further white settlement.

Homestead farmers moved into the valley, settling initially in the fertile Clark's Fork Bottom near today's Billings. In 1877, businessman Perry W. McAdow established a sawmill and cropland there to provision the nearby Crow agency. As fate would have it, that September the Nez Perce, on their epic flight from the U.S. Army, reached the bottomland and burned many cabins, along with McAdow's store and saloon; the sawmill was saved. The community rebuilt, and McAdow named the revived town Coulson. Once again, crops began pushing their way up through the alluvial soils. A bountiful harvest the following year included squash, beets, turnips, and onions along with more exotic fare—watermelon, corn "10 to 12 inches long in the ear," enormous tomatoes, and cucumbers "half an arm in length."[4]

McAdow dug the first irrigation canal in the region as he moved from farming and flour milling in the Gallatin Valley to stake his fortune in the Yellowstone Valley. His large farm on Clark's Fork Bottom included an impressive barn, "an immense one, the posts being 12 inches square and the frame being built like a quartz mill."[5]

Soon, more log cabins and barns appeared, and when the Northern Pacific Railroad came through in 1882, the city of Billings took hold. The Yellowstone Basin proved an expansive farming area, and the resulting barns in the upper Yellowstone canyons and valleys are a mixed lot, ranging from log sheds and stables of the late nineteenth century to Gothic arch barns of the 1920s. Near its headwaters, the Yellowstone is a powerful presence in the Paradise Valley, and its river-worn cobblestones inspired some massive and masterful stone foundations.

Lower down, the river is swollen by the waters of the Boulder River and the Clark's Fork of the Yellowstone that cascade off the Beartooth Plateau and tumble down from the Absaroka and Pryor mountains. In Carbon and Stillwater counties, there are small communities founded in the alpine valleys that mix mining and agriculture. Today, the barns of Carbon County reveal much about the rural enclaves of miners and ethnic settlers who built with stone and log. In addition, the emerging technology of the twentieth century can be found here, as illustrated in Gothic-arched barns and the concrete barn at Fromberg, a barn prototype that is most uncommon in Montana (see Gebo Barn, page 242).

By the 1870s, cattle driven from points south and west had started to fill eastern Montana's prairies. Miles City photographer L. A. Huffman caught this herd crossing the Powder River at the "old N– crossing." The N Bar Ranch is on the far bank.

Eastern Montana

Beyond Billings and Coulson lie the grasslands of eastern Montana, part of the Great Plains "empire of grass," as historian Walter Prescott Webb called it, and one of the last regions in the West to be settled. By the late 1870s, with the buffalo fast disappearing, the Indian wars drawing to a close, and ranges in territories to the south filling up, cattlemen began to fan out onto the plains of southeastern Montana, seeking open range for their herds and a shortcut to eastern markets.

The area did not take long to develop. The *Rocky Mountain Husbandman* took stock of the land rush in December 1879, declaring: "Eastern Montana is booming. The shackles that have bound it in years past have suddenly burst asunder and its latent resources are beginning to be aroused and developed. . . . Stock is pouring in from every hand; farmers are locating land, and the mountains are alive with prospectors."[6] Two years later came a rush of settlers who had read, or heard of, General James Brisbin's book *The Beef Bonanza; or How to Get Rich on the Plains*, which declared that "Montana has undoubtedly the best grazing grounds in America. . . . The Yellowstone, Big Horn, Tongue River and Powder River regions contain the maximum advantages to the cattle-grower."[7]

Until 1881, the completion of the Northern Pacific Railroad was stalled at Bismarck, North Dakota, on Montana's doorstep. Prior to arrival of the railroad, eastern Montana's newly opened ranges offered an alternative route to take cattle to Union Pacific railheads to the south. In 1879, ranchers from western Montana blazed a shortcut that became their "eastern route." They drove their herds out of Deer Lodge to the Sun River Valley, on along the Missouri and Musselshell rivers, then across the Yellowstone basin and south by way of the Tongue and Powder rivers and Rosebud Creek. From there, it was a straight shot along the North Platte to Fort Laramie and western Nebraska.[8]

For a couple of years, this "eastern route" served as a cattle highway as the captains of Montana's beef industry—including Conrad Kohrs and John Bielenberg, Con Warren, and Charles Conrad—and English and Scottish cattle syndicates drove thousands of head toward the railheads. Even as they did, the long-awaited rail connection to Montana pushed on across the North Dakota border to what was then "Miles Town." Con Warren later described the 1880 drive, the last one prior to the completion of the Northern Pacific, noting that "they summered on the Tongue and shipped for the first time on the Northern Pacific from Miles City in 1881."[9]

Miles City began as an army cantonment at the confluence of the Tongue and Yellowstone rivers. That temporary station grew to become Fort Keogh in 1876. The coming of the Northern Pacific positioned Miles City at the center of a giant stock range, with cattle and horses grazing a hundred miles in every direction and little to get in their way. For several high-flying, freewheeling years, cattle piled into eastern Montana. The land, it seemed, was limitless.

And so the big sky and Montana's eastern plains became the backdrop to legendary cattle drives, the route up from Texas heavily traveled by some of the biggest outfits ever to move beef on the hoof. The XIT, the Matador, the Powder River Cattle Company—all discovered the protein-rich grasses of eastern Montana and began to fatten their livestock here. The numbers were staggering: in 1882 E. S. "Zeke" and Henry H. J. Newman's Niobrara Cattle Company trailed twelve thousand cattle into Montana's Powder River Valley. (Their N Bar cattle also grazed at the mouth of the Musselshell River.) The Scott and Hanks Company of Idaho drove twenty thousand head into southeastern Montana, establishing a home ranch and the SH horse ranch on the Tongue and Powder rivers.

In its earliest years, the Scott and Hanks Company home ranch on the Powder River included these tidy log cottages with beautiful stone fireplaces.

The Tongue River range quickly filled with cattle, as the *Yellowstone Journal* out of Miles City reported in 1884: "Mr. Scott has just returned from Nevada where he has purchased 5,000 head of fine stock cattle which will be driven to his range on upper Tongue River as soon as the weather opens up."[10]

The early ranchers built long, spare buildings—a low-slung, high plains building form that seemed to sprout from the prairie soil. With thick walls of native sandstone or log and low pole roofs covered with sod or dirt and gravel, the architecture provided shelter from the scorching summer sun and the cold, blustery winters of the arid West. The people of the plains of Colorado, Wyoming, and Montana remember these buildings: houses and barns formed of a chain of cribs that hunkered down into the landscape and could be added on to over the seasons. The OW Ranch, with its exceptional multiple-crib log barn, is a classic assembly of such barn buildings (see page 260).

Dugout buildings were also part of the built environment in eastern Montana. Most are gone, but the very few that remain speak volumes about the experience of carving out an agricultural existence in a semidesert environment.

Eastern Montana was also good sheep range. When Fort Keogh was established, it had been three-quarters of a century since Lewis and Clark spotted a "gang of 40 Big Horn sheep" near Pompeys Pillar. In 1876, with troops to provision, General Nelson Miles convinced sheep rancher John Burgess, who was en route from California to the Black Hills with a large flock, to stop instead near the new fort. The following year, the general's nephew, George Miles, and his partner, Captain Frank Baldwin, bought out Burgess.[11] Starting with a flock of eighteen hundred rams and ewes, they provided mutton to the fort and shipped the wool clip to Boston. Baldwin later settled on Rosebud Creek, where his ranch remains in the hands of his descendents.

And on Pumpkin Creek, sheep king Robert R. Selway ran the largest operation in eastern Montana, with a hundred bands of sheep each numbering between two thousand and three thousand head.[12] Sheep were less than popular with some cattlemen in the region, however. In 1900, the only noteworthy range-war incident in Montana's history took place near Birney when three thousand of Selway's sheep in a winter camp were clubbed to death by a dozen masked cowboys while one cowboy held a rifle on the sheepherder.

The Tongue River and Powder River Basins

Among the early ranchers in eastern Montana were cattlemen like Captain Albert Brown, whose cowboys moved cattle across such an extensive range that they would leave home in May and aim to return by Thanksgiving. Claiming twenty thousand to thirty thousand head early on, the Brown Cattle Company maintained its herds throughout the Tongue and Powder River valleys and beyond. Family members still tell of early years when cattle drank from the Platte River in Nebraska in spring and shipped out to market in fall on the Yellowstone River, some four hundred miles away.

Carrying capacity of the ranches in these valleys came down to two factors: water and grass. Southeastern Montana has a dry climate, averaging thirteen or fourteen inches of rainfall each year. The most successful ranches were those that could irrigate and grow enough hay for supplemental feeding of the cattle and that found springs and developed stock tanks out on the range so that cattle could be moved through the year to fresh pastures with good grass.

In addition to cattle, horse raising was big business on the ranges of eastern Montana. In a time when horsepower moved the world, draft horses and mounts for the military were always needed. Percherons and Belgians for working in the fields and hauling heavy loads and Thoroughbreds for cavalry regiments were primary exports from Montana's stock raisers to eastern and foreign markets. During Britain's Boer War, from 1899 to 1902, Billings and Miles City became hubs for international horse sales and shipping. This market was boosted by British and Scottish stockmen such as Malcolm Moncreiffe and Oliver H. Wallop, who established Thoroughbred horse ranches on the Tongue River and Otter Creek, respectively, and secured lucrative contracts for the breeding and marketing of five-year-old horses, selected for their solid coloring and sixteen-hand conformation.[13]

In 1909, Fort Keogh was commissioned as an army remount station, and until the end of World War I the post supplied horses to the U.S., French, and British troops fighting in Europe. To secure quality stock, the army supplied Thoroughbred stallions to ranchers in the region, who in turn bred them with their range mares to produce the desired sixteen-hand mounts for the military. Here in southeastern Montana, a part of the state best suited to livestock grazing, the horse industry thrived.

Ethnic settlers and homesteaders were also drawn to the region, particularly following passage of the Enlarged Homestead Act in 1909. Among the immigrants was a sizeable Norwegian community near Birney. These newcomers brought exceptional masonry and carpentry skills. In a region of limited timber and building logs, stone became an important part of the building stock for houses, barns, and chicken coops.

Farming in the Region

The completion of the Northern Pacific Railroad to Fort Keogh and Miles City in 1881 opened eastern Montana to farming and settlement. In the arid regions, farming was helped along in the early twentieth century by the development of the Huntley Irrigation Project and Lower Yellowstone Irrigation Project. These irrigation works served many young communities, enabling the cultivation of benchlands that formerly supported grazing only. At Huntley, a dryland demonstration farm was established that is now the Southern Montana Agricultural Research Center of Montana State University. Montana's sugar beet industry, made possible by irrigation, began in 1906 with the Billings Sugar Company, spread throughout the sandy-loamed Yellowstone River country, and eventually led to large processing plants run by Holly Sugar Company in the 1930s at Hardin and Sidney.

Out on the eastern edges of Montana, the environs were much like those of the Dakotas, and many settlers drawn to Glendive and Sidney were farmers who had moved into the Dakotas and then simply pushed a little farther west to claim some of the last of the homestead lands available in Montana, and in the United States. Charles Buller, for example, was a Mennonite whose Russian father had moved the family to South Dakota while looking for a permanent place to settle. A Mennonite newspaper convinced them to move to the Glendive area, where they finally located a permanent home, only a few decades and thirty-five moves after they had begun their quest.

The Yellowstone Basin was then, and is still, big wheat and cattle country. The landscape is a complex network of county roads edging around the enormous wheat fields that are interspersed with livestock pastures. Farms dot these rolling prairies, and their red, gambrel-roofed barns are visible across long distances. In the Mon-Dak region, there are still many early-twentieth-century barns, built to house dairy cattle, livestock feed, and bumper harvests of the prairies.

One of the area's most prominent farmers was Thomas D. Campbell. Inspired by the bonanza farms of the 1880s, Campbell, in 1918, combined land leased from the Crow tribe with land he owned to create a farm that encompassed 150,000 acres. It was, literally, the biggest wheat farm the world had ever seen. With five hundred plows and thirty-two harvesting

The Bones Brothers (left to right), "Big Bones" Allen Roberts Alderson, Floyd Alderson (kneeling), and "Little Bones" Irving Alderson, circa 1920s

combines and threshers, this farm produced wheat to feed the world during World War I. In the 1930s, Campbell acted as a consultant to the large farming collectives of the Socialist era in Russia.[14]

Dudes at the Bones Brothers Ranch

Dude Ranching

Following World War I, many stock raisers found themselves struggling financially, and the stage was set for an expansion of the informal practice of taking in paying guests on western ranches. Dude ranching capitalized on Americans' fascination with the Old West, and western entrepreneurs seized upon the opportunity to attract paying guests and thereby stay in business.

The first dude ranchers in the United States were the Eaton brothers—Alden, Willis, and Howard—who established a horse and cattle ranch near the North Dakota badlands in 1879. The Eatons hosted their first paying guest in 1882; among their visitors was an enthusiastic Theodore Roosevelt, who was drawn to buffalo hunting and cow punching.

Irving Alderson Jr. (right) with friend Jimmy Kelso; Santa, the ranch St. Bernard; and horses Strawberry Shortcake (left) and Frannie, circa 1930s

The Eatons moved in 1902 to the base of the Bighorn Mountains in Wyoming, and nearby Sheridan became the epicenter of the industry. As Bucky King wrote in the *Dude Ranch Connection*, "The one-hundred-mile area between Buffalo, Sheridan and Birney, Montana was the Dude Capital of the West during the twenties and thirties." The Dude Ranchers' Association was formed in 1926 in Bozeman, Montana. At the time, the local newspaper noted, "Dude ranching as a business had grown from a boarding house proposition with an occasional summer visitor into an industry that had brought an estimated revenue of $1,000,000 to Montana within the present year."[15]

During the 1920s, the Bones Brothers Ranch, the Quarter Circle U, and other Tongue River ranches took up dude ranching, building guest cabins and tents, dining rooms, and horse barns set up to accommodate guests. From 1923 until 1963, the Bones Brothers Ranch offered a working vacation to the sons and daughters of eastern families, with horseback riding, roping, branding, rodeos with the Northern Cheyenne, baseball, fishing, dancing with cowboys, and swimming in the Tongue River.[16]

• • •

AS JOHN WESLEY POWELL reported in 1878, it is somewhere around the 100th meridian where rainfall tapers off and the earth runs flat, that the High Plains and the West begin. This West is a land of empty sky and powerful weather. It is at about the 104th meridian that Montana begins, watered by distant mountains that rise like a mirage off the prairie floor. Here there are basins that capture the clear mountain waters of the Bighorns, the Crazies, and the Snowies and deliver them into the Yellowstone River. In turn, the Yellowstone pours into the Missouri at Fort Union just before it makes its exit into the Dakotas.

If western Montana is the place where ranching and farming first took hold in our state, eastern Montana is the place where the raisers and the growers still hold fast. Agriculture is the main event in all places east of Lewistown. This is country where the Old West still exists, where cowboys work cattle, where neighbors hold roundups, where children are raised to work the land, and where people still wake up with one ear on the early-morning farm report.

This, too, is still Indian country, with sprawling reservations that are home to the Sioux, the Assiniboine, the Northern Cheyenne, and the Crow Nations—people who continue to practice traditions that have been kept alive over countless generations. This is country where no matter where you walk in any direction, given a half mile or so, you are bound to stumble across an archeological site yielding the glint of a flint knife or a chert scraper that serves as the reminder of an earlier day and of a time when people and grazing animals had a different understanding of each other.

Bonnell Barn

PARK COUNTY, 1922

When Oliver and Lucy Bonnell built their Gothic arch–roofed barn in 1922, they were undoubtedly influenced by promotional literature on farming practices being circulated to rural people during the 1920s and 1930s by government agencies. The barn illustrates the design features recommended for dairy barns, including rows of windows for good lighting, cupolas for ventilation, ample loft space for storage, and concrete floors with gutters for sanitation. The Gothic arch, or "rainbow arch," roof represents the culmination of evolving barn design, as roofs grew from the simple gable to the popular gambrel to the impressive and spacious Gothic style.

For decades, dairying in Montana was largely limited to individual farmers who kept a herd of cows to supply milk and cream for home use. After World War I, however, dairying became a sideline on many farms to supplement the family income, and by 1925 Montana had 228,000 cows that kept seventy-four creameries and nine cheese factories in business. The largest cheese factory in Montana was located close to the Bonnell Ranch, northwest of Wilsall. Clyde Park also had a creamery as early as 1908.

The Bonnells and their successors sold their milk to several dairies in the area, including the Windsor Dairy, the Yellowstone Dairy, and the Farmer's Creamery. When the Farmer's Creamery got the contract to supply dairy products to all of the government concessions and private companies in Yellowstone National Park in 1928, it became the primary customer for dairies in the area.

Miller Ranch Barn

PARK COUNTY, CIRCA 1890

The barn on the Miller Ranch is an excellent example of a late-nineteenth-century barn, constructed when milled lumber and rail-shipped building supplies were readily available. The barn was likely built in the late 1890s. A Montana pioneer, Edward Miller came to the territory in 1865 and cowboyed as a teenager in the Gallatin and Paradise valleys when the area was largely open range. The land north of the Yellowstone River along the Shields River was surveyed by the General Land Office in 1873 in order to facilitate land sales, but it was not until after the Northern Pacific Railroad came through in 1882 and established the rail hub of Livingston that settlement really picked up. The railroad drew a steady stream of homesteaders who claimed acreage under the 1862 Homestead Act or purchased property from the Northern Pacific. Miller likely filed for his homestead in 1894. After five years of improving his land, he received a patent for the land in 1899.

Very early barns in this area were simple one-story log structures built from logs hand-hewn with an axe and adze and often put together without the aid of nails. As the railroad brought more people, it also brought lumberyards and mills and professional building contractors. The Miller Barn, for example, is framed with saw-cut lumber and sided with saw-cut boards and battens. Each piece is fastened together with machine-made nails. This type of construction, of course, was common from the 1890s through the 1940s, but for many years when Montana was a territory, its pioneers could only dream of having building materials at the ready.

The current owners have carefully repaired the barn, honoring the honesty of its construction and the beautiful aesthetic of its unpainted exterior.

Wampler Barn

PARK COUNTY, 1912

Beginning in the 1890s, Croatian and Italian stonemasons arrived in Montana to find good work building schools, courthouses, commercial buildings, and, at the head of the Paradise Valley, the hotels and buildings of Yellowstone National Park. After work slowed in the park and the bustling towns, many masons found work in less populated places nearby.

In the area between Emigrant and Livingston, Italian stonemasons constructed a number of barns between 1910 and 1920. During those years, they would collect stone all winter long and then build barns, houses, and outbuildings during the summer.

Three generations of the Shimmin family: Kory, Bruce, and owner-operator Riley

John Taze Wampler and his wife, Margaret "Bell" Wampler, settled near Pray in 1901. The son of a Virginia farmer, Wampler was no stranger to hard work and was soon running a successful farm. Misfortune struck in 1911 when his barn burned down, but he lost no time in rebounding, hiring local stonemasons to build a new barn the following year.

The cobblestone masonry is particularly distinctive in this part of the upper Yellowstone Valley where the river shaped the landscape and the river rock remains in abundance. With its thick stone walls and round roof, the Wampler Barn is strikingly similar to the barns at Chico Hot Springs and at the Slip-N-Slide Ranch near Emigrant (see page 233), which all combined traditional stonework on the lower level and progressive twentieth-century loft design above to produce exceptionally sturdy, efficient barns.

Malcolm Barn

PARK COUNTY, 1918

The Malcolm Barn is one of several in the Paradise Valley built from the local river rock that is a hallmark of the upper Yellowstone drainage. While some of these barns were built by Swiss-Italian stonemasons, this one appears to have been built by its owner, Alec Malcolm, and sons in 1918. Most of the rock was easily quarried from the cliffs of the Absaroka Range, where Precambrian rock lies exposed under volcanic rocks left from an eruption about 50 million years ago.

The stone walls are laid in an irregular, or rubble-stone, pattern perpendicular to the hillside, and they extend to either side to form retaining walls. Interestingly, the stone walls do not continue across the façade; instead the Malcolms used wood to build the front and frame the wide central-alleyway doors.

The Malcolms used the barn for milk cows and hay and grain storage. At one time, they housed pigs in a north wing, which was identical to the existing milking shed on the south. Unfortunately, the pig shed was destroyed by a heavy snowfall in 1965. The barn features a very gentle gambrel roof that frames the large Bench K brand used by current owners Joseph and Millicent Kimbro.

Rigler Barn

PARK COUNTY, 1912–1915

Paradise Valley is home to a number of well-crafted stone barns, reflecting a high level of building skill in the community and a ready supply of stone for building. The story of the Rigler family is told in one of those stone barns. Paul Rigler, a stonemason from Austria, would have been proud of his wife, Cecilia, and his six sons for building a stone barn in his adopted country, had he lived to see that barn. In 1893, Rigler came to Montana to work in the mines at Cokedale; he then married and ran a saloon and store in Aldridge for several years. In 1908, the Riglers bought a ranch in the Paradise Valley, but Paul died in 1909, leaving his family with a twenty-five-hundred-dollar mortgage.

Cecilia tried to sell the ranch and move on, but the buyer died before paying her, so she stayed on with her six sons and worked hard to make ends meet. Taking advantage of their proximity to Yellowstone National Park, they raised cabbages, potatoes, and apples to sell at Corwin Springs; took in boarders; and transported park travelers to the train station. The boys took various jobs in the park; in 1914, for example, Tony and Charlie Rigler bought horses to pull a water truck through Yellowstone, spraying the dirt roads to keep dust down for the tourists.

The Riglers put their hard-earned money back into the ranch, and, beginning in 1912, the entire family, including Cecilia, worked together to collect stone from the nearby hills to build a barn for their horses and tack. They hired a carpenter who had done work in Yellowstone to frame the barn and a local stonemason to lay the two-foot-thick walls. By 1915, the barn was complete and the mortgage paid in full. The Riglers' sweat equity paid off, and they were later able to expand the original ranch, acquiring what they called "the Upper Place" in 1919 and the Old Joe Brown Ranch on Joe Brown and Slip-N-Slide creeks in 1920. From 1920 to 1966, the ranch was known as the Rigler Ranch; since 1966, it has been called the Slip-N-Slide Ranch.

The Rigler great-grandchildren still operate the ranch; they raise cattle and welcome guests for horseback excursions. The Rigler story is a Montana saga of determination, hardship, and perseverance, the kind that is deeply woven into the fabric of the state's culture.

Fisher Farmstead Barn

PARK COUNTY, 1926

The Fisher Dairy Barn sits perched atop the Cottonwood Bench in a perfect state of preservation. The barn was built in 1926 by Lewis A. Fisher, his son Alfred, and his neighbors at a cost of $5,000. Lewis Fisher had been a pharmacist in Fayette County, Iowa, and brought his family to the Clyde Park area in 1914, when he was close to fifty years old. State promotional reports of the time encouraged diversified farms, and many dairies were launched in the Gallatin and Shields valleys to supply local creameries in the 1920s. Agricultural experiment station bulletins provided guidance on dairy management and extolled the virtues of the Gothic-roof barn for its substantial, unobstructed hayloft. These bulletins also recommended designs for comfort and sanitation, featuring rows of windows in every stall, concrete floors, waste gutters, and track systems for feeding and cleaning, all of which were put to good use in this barn.

The Fishers ran the dairy for a number of years, but like many ranchers and farmers across the state, they lost the farm during the Great Depression. The barn was not used as a dairy again. Even so, the barn has never been altered in the years since. From the beautiful crystal-ball weathervanes atop the wood-shingle roof down to the milking stanchions and the humble manure bucket, the Fisher Barn remains in pristine condition.

The interior of the barn is so well preserved, it appears as though the cows might show up anytime for milking. Milking stanchions line the left wall, and the manure bucket and track run down the center aisle.

Cobb Barns

PARK COUNTY, 1890s

(above) The horse barn stands in the foreground, and the granary is across the road.

(opposite bottom) Horse barn, side view

Jerome (Jerry) Cobb, a Missouri native and Park County pioneer, arrived in 1880 and purchased land in Paradise Valley, first south of Livingston in the 1890s, then, shortly thereafter, this Suce Creek homestead.

Cobb's horse barn and granary are good examples of two different early-day log buildings that may have been put up single-handedly by Cobb or with the help of a carpenter or neighbors. The log work on the horse barn is a distinctive example of the *piece-sur-piece* construction often favored by French-Canadian carpenters, who stacked short lengths of logs between two channel posts to form the walls. In this case, the walls have such large spaces between the stacked logs that short and skinny pieces of logs were used as chinking, with no mortar or daubing to cover them.

Granary, viewed from back

The granary across the road is a well-built late-1800s barn. The logs are all of uniform size, the lapped corners are notched and stacked quite neatly, and the gable ends are shingled, a finish that is rarely seen on a log barn and may have been a later addition. These elements suggest that the builder was likely a professional or at least quite skilled at his trade. For many years, Cobb was the threshing man in the valley, doing much of the custom harvesting in the surrounding area. Housing the horses and feed for his threshing team, his barns were essential components of his successful business.

Hoppe/Reed Barn

PARK COUNTY, CIRCA 1906

The B Bar Ranch occupies exceptional land in the Tom Miner Basin that for centuries was prized as a rich hunting ground by the Crows and the Mountain Shoshones and, later, by fur trappers and early-day miners. The basin got its name in the mid-1860s when nearby Emigrant Gulch was a lively gold-mining camp and Thomas J. Miner began trapping in the basin.

After homesteading opened in the late 1890s, the basin quickly became dotted with log cabins and barns. Hugo B. Hoppe, son of pioneer Hugo J. Hoppe, claimed land in 1906 and built the barn shortly thereafter. It is a wonderful early-1900s draft horse barn, built with heavy timbers, wide stalls, meaty floorboards, and a spacious hayloft. Allen Reed bought the Hoppe place in 1922 and worked the ranch with his family for more than twenty years.

The Reed place is now part of the B Bar Ranch, a working guest ranch. The ranch proudly raises Suffolk Punch draft horses, manages a herd of Ancient White Park cattle (and sells organic, grass-finished beef throughout Montana), and operates a high-altitude greenhouse and organic garden featuring heirloom vegetables and herbs. The barn continues to be important in the ranch's operations, as draft horses are used almost daily to plow, hay, and haul manure and heavy equipment. In 2001, owner Maryanne Mott restored the barn inside and out with keen sensitivity to its original character. Today, the B Bar Barn loft is full of hay to feed the workhorses that are still much a part of everyday life on this ranch.

(below left) Draft horses and foreman Don Yerian

(below center) A conveyor track moves hay into the loft for stacking.

The barns on the Shiplet Ranch represent two different eras of ranching in the Shields Valley.

Shiplet Barns

PARK COUNTY, CIRCA 1880, 1921

Jacky Shiplet and her grandson Steven

The barns on the Shiplet Ranch tell a long story of livestock raising on this historic Clyde Park property. During the open-range days, the Shields Valley saw thousands of Texas cattle arrive to graze the rich grasses of its bottomlands. The log buildings on the property date to the 1870s or early 1880s. A cabin-barn made of rounded logs was probably built first, followed by a second barn of square-hewn logs.

John Harvey opened a post office on his ranch in 1887. Allegedly, the name Clyde Park was chosen because he had just imported a registered Clydesdale stallion from England, and the name stuck when the post office was moved to the present site of Clyde Park in 1901. By that time, Harvey and Tom Tregloan ran one of the largest cattle outfits in the valley.

In 1909, when a branch line of the Northern Pacific came through, railroad workers accidently burned one of the original barns on the ranch. The railroad paid for the present red barn to be built, larger in scale, with a gambrel roof and design features of a proud twentieth-century barn.

In the 1930s, Alban and Howard Bodine purchased the ranch. For a time, Alban Bodine worked for James J. Hill, owner of the Great Northern Railway, and in 1919 he had moved to Livingston to manage the properties of Hill's youngest son, Walter. Bodine later went into ranching on his own, and his Bodine Herefords won championship awards throughout the West.

In 1956, the ranch sold again, this time to Robert (Bob) Shiplet, a professional rodeo cowboy, who, from 1957 to 1961, won many honors at rodeos throughout the Northwest. The Shiplets raised prize-winning Black Angus cattle. The bull painted on the gable end of the barn bears the Shiplet brand, which is the state of Montana.

Stovall Barn

CARBON COUNTY, 1958

The Stovall Barn stands majestic against the foot of the Beartooth Mountains near Luther. This barn is young compared with many in the area, having been built in 1958 by John Durrin for rancher Robert Stovall. It is a symbol of the years following World War II when Montana stockmen enjoyed prosperous times. The barn was considered one of the finest in the county when it was built, and since then, it has been meticulously maintained. In addition to five double horse stalls, a tack room, and one stud stall, it has a heated milking parlor with four milking stanchions, a concrete alley, electricity, and running water. Barn dances were a special once-a-year event enjoyed by all.

Today, the barn is part of the Switchback Ranch, an extensive cattle ranch in the shadow of the Beartooth Mountains. The barn, which mostly houses ATVs and other modern equipment, is a wonderful example of a building that has been adapted over the years to changing technology and ranching methods.

A modern use for horse stalls

Gebo Barn

CARBON COUNTY, 1909

The Gebo Barn was the pride of wealthy coal-mine developer Samuel W. (Sam) Gebo, who built this barn at Fromberg between 1907 and 1909. The massive concrete barn was built on a grand scale to house purebred Belgian horses, other livestock, and farm equipment on what was then the largest ranch in Carbon County. A local reporter described it as "one of the finest farms in Montana, equipped from the woven wire fence right into the fireplace with all the modern implements and accessories."

When the barn was completed in 1909, Gebo christened it with a lavish dance in the cavernous loft. Refreshments

and a full orchestra were among the entertainments. Three hundred people from around the state braved bad weather and treacherous roads to attend the celebration. Soon after, Gebo announced in the *Bridger Times* that his registered stud, Monico, would be "standing" at the Fromberg Ranch and other locations all summer. The stud fee was twenty dollars.

Some details have been lost to time, but it is likely that Gebo chose concrete as the primary building material for the barn because of its ready availability from the newly opened Gibson Concrete Works in Fromberg. Concrete residue on the roof rafters and sheathing indicates that the lumber was first used to construct forms for the barn's poured-concrete walls, then recycled to build the roof.

Concrete was an increasingly important twentieth-century building material, and between 1905 and 1910 two of the leading Portland cement manufacturers published full-length books on the benefits and uses of concrete for farm buildings. These books, and the popular plan books of the day, promoted the use of concrete for everything from troughs to foundations and retaining walls to complete buildings. The walls of the Gebo Barn were an efficient and fireproof choice; they also represented the latest thinking on sanitary engineering since they could be washed down completely without fear of rot or pest infestation.

Concrete barns never became very popular in Montana or elsewhere. Though concrete manufacturers boasted their many benefits, animals did not thrive in them given that the barns stayed chilly and damp inside year-round.

The interior has been remodeled for use as a restaurant.

Kent Dairy Round Barn

CARBON COUNTY, 1939–1940

For more than sixty years, the Kent Dairy Round Barn has marked the north entrance to Red Lodge. It has the distinction of being the only round brick barn in Montana and one of the few brick barns in the state. Built by Finnish immigrants under the supervision of master barn builder Emery McNamee, the barn is an important pre–World War II agricultural building in the Rocky Mountain region.

Ephraim Kentta (anglicized to Kent) brought his family from Finland to Montana in the early 1900s and found work in the coal mines near Red Lodge. His wife, Fiina, started a small dairy to earn extra money, and by the 1930s she had many cows. To expand the operation, Ephraim and his sons Armas, Harry, James, Leo, and Waino built a new barn between 1939 and 1940 with red bricks salvaged from a nearby abandoned building.

The Kent family painstakingly cleaned the mortar from each brick before setting it in the barn walls. Ephraim and Harry together hand-hewed two massive cottonwood beams to support the main-floor ceiling, and the Kent brothers shaped each of the thirty-seven support posts on the first floor with a drawknife. It was a labor of love and learning as none of the Kents had any prior training as carpenter or mason.

Fortunately for them, the Kents did have one of the country's foremost round-barn contractors on-site to help them. Emery McNamee and his father, Isaac, were among a group of well-known contractors in Indiana who built the first true round barns in that state from 1900 through the 1920s. The McNamees started out in Indiana and built their way from Iowa to South Dakota and as far as Saskatchewan. At age eighty-one, Emery was living in Red Lodge and guided the construction of the Kent barn. Although McNamee lived to be one hundred years old, round barns went out of style by the mid-twentieth century, and it is assumed that the Kent Barn was the last of McNamee's large-scale projects.

Renlund Barn

CARBON COUNTY, CIRCA 1909

Built with meticulous craftsmanship to stand for a century, the Renlund Barn represents the finest of Finnish log-building traditions. With its spacious rectangular plan, full-height log walls, steep gable roof, and beautiful dovetail-notched corners, the barn is a model from which to understand other regional Finnish barns.

Sanfrid Renlund built the barn to high standards, likely with the help of other Finnish immigrants who lived nearby. Renlund immigrated from Finland in 1890, and for a number of years he ran a general merchandise store and later a butcher shop in Red Lodge. In 1909, he bought a 160-acre homestead claim in Red Lodge from W. F. Meyer, a prominent Red Lodge banker. The barn was built shortly after Renlund purchased the land and likely housed a herd of milk cows, a few horses, and hay and grain. Renlund retired in 1930 and sold the ranch to his daughter and son-in-law, who kept it in the family until 1994.

Large numbers of Finnish immigrants came to Montana beginning in the 1890s, and many were drawn to the coal-mining town of Red Lodge. "The Finns," as they became known, were one of the first ethnic groups to arrive, but as the availability of jobs in the mines fluctuated, many relied on their native trade—carpentry. They settled the section of Red Lodge known as Finn Town and built many homes as well as such larger community buildings as boardinghouses, saunas, stores, a church, the Workers Hall, Fox Hall, and the Opera House. While many of these in-town buildings no longer remain or have been remodeled extensively, the Renlund Barn stands unchanged, a legacy to the history, character, and craft of Red Lodge's earliest Finnish residents.

Vincent/Long Barn

CARBON COUNTY, 1930

This simple, three-bay log horse barn, standing in the high-alpine region of the Beartooth Mountains, has seen its share of snow over the decades. Constructed of small-diameter lodgepole pine logs from the surrounding forest, the barn consists of a narrow center aisle flanked by larger rooms with double horse stalls on each side. At one time, there was also a full-length milking shed on the north side.

The barn most likely was built by James E. Stewart, who homesteaded the land in 1911, or by Francis Vincent, who bought the property in 1913 and raised hay, hogs, milk cows, horses, and chickens. A large number of Vincent's neighbors were from Norway and Sweden. Many of them found this area to be similar to the mountainous Scandinavian and Swiss regions in Europe and well suited to dairying.

Rankin Barn

CARBON COUNTY, CIRCA 1916

Located at the base of the Red Lodge Creek plateau in the Beartooth Mountains, the Rankin Barn represents the ingenious, sturdy farm structures built by young families during the homestead era. The barn was most likely built by Lawrence Rankin, who came to Montana from Missouri in 1916. He purchased 160 acres near Luther from homesteader William Wilson, who had held the property for only one year. Rankin, who raised four children with his wife, Nora, survived the tough years of the Great Depression on this land and ranched here until his death in 1940.

The remote location and the layout of the interior of the Rankin Barn suggest that it was a calving barn and shelter for range cattle. To build the barn, Rankin gathered stones from the nearby creek bed for the foundation and cut small lodge-pole pines for the walls. He then peeled and steeple-notched the log ends and, with a helper, stacked the logs to form walls. A shallow loft made it possible to store hay and feed.

Nelson/Wegner Barn

CARBON COUNTY, CIRCA 1903

The craftsmanship and building methods of the Nelson/Wegner Barn suggest a build date after 1903, when John A. Nelson purchased the ranch. Nelson came from Norway to the United States as a baby in 1868 and grew up in Minnesota and South Dakota. In 1896, he and his wife, parents, and sister settled near Bridger. In 1903, he bought 160 acres on Clark's Fork of the Yellowstone River and later patented more than three hundred acres adjoining his ranch.

While the name of the carpenter who built the barn is unknown, the neat and even mortar joints and coursed rock-faced sandstone indicate that a skilled stonemason completed the job. The barn was built of local sandstone, which likely was shipped by rail from nearby quarries in Laurel or Columbus. About 1900, quarrymen east of the Rocky Mountains began to excavate large quantities of Cretaceous sandstone. By 1908, this stone was profitably being quarried in eight counties, and practically every town from Wibaux to Livingston had a sandstone quarry. Columbus became the principal area in Montana for sandstone production in terms of both quality and quantity, and stone for the state capitol was quarried there.

Skilled stonemasons arrived in Montana with other immigrants who had passed through Ellis Island on their way to the West in the early 1900s. These men came primarily from small villages in the Italian Alps and Austria where stone masonry had thrived as a proud industry for many generations. While some of these masons engaged in mine work in Montana, the abundance of stone and the "building boom" in newborn communities encouraged many to continue their trade. The largest number of stonemasons in Montana settled near Lewistown, but Carbon County had its fair share. Today, the many impressive stone buildings that remain in the region are the lasting testament to their skills.

Southern Montana Agricultural Research Center Barns, Huntley

YELLOWSTONE COUNTY, MID-1920s

When the Huntley Irrigation Project opened to settlers in 1907, many of the newcomers were not farmers by trade, and few had experience with irrigation. To introduce homesteaders to irrigated farming, supervising engineer H. N. Savage negotiated with the Montana Agricultural Experiment Station in Bozeman to establish a demonstration farm on the Osborn town site, then an undeveloped railroad siding on the Northern Pacific Railroad.

Savage enlisted the support of irrigation expert I. D. O'Donnell of Hesper Farm and the U.S. Bureau of Reclamation. O'Donnell accepted the challenge, and the Bureau of Reclamation launched "The Farm" in 1907. Three years later, it evolved into a U.S. Department of Agriculture Bureau of Plant Industries Field Station, cooperating with Montana State College and the Montana Agricultural Experiment Station. The large dairy barns, silos, and horse barns were built in the mid-1920s. Models of innovation at the time, they were used to demonstrate best practices in dairy farming. The experiment station served in this capacity for three decades. In 1961, the animals were moved to Ogden, Utah.

Hesper Farm Barn

YELLOWSTONE COUNTY, 1884

The barn on the Hesper Farm west of Billings is linked to the early development of the city of Billings. The barn and a house were constructed in 1884 on 160 acres under direction of the Reverend B. F. Shuart, an agent for Frederick Billings, the city's namesake, just after the town was founded. The barn is a wonderful 1880s gable-roof barn with side wings. It housed both dairy cows and draft horses. Later, from the mid-1940s to the 1980s, it was used as a turkey barn.

Shuart named the property Hesper after the evening star Hesperus, king of the western lands in Greek mythology. The Reverend Shuart and Hesper Farm were both key in the development of farming in Yellowstone County. Shuart planted the first alfalfa here in 1884, ran a dairy that was the first to sell butter in Billings, and developed an early irrigation system on the farm. In 1892, Hesper Farm was sold to Billings's nephew, E. G. Bailey, and I. D. O'Donnell, manager of Billings's three-thousand-acre ranch and the Big Ditch irrigation project.

After purchasing Hesper Farm, O'Donnell expanded on Shuart's work and became one of Montana's best-known agricultural leaders. He popularized the raising of such crops as alfalfa and sugar beets and promoted irrigation and reclamation on a national scale. In 1919, the *Country Gentlemen* named him "Best Farmer in Montana," and in 1925 the *Saturday Evening Post* published a feature about him entitled "Montana's Foremost Farmer." Hesper Farm became a demonstration site for O'Donnell, and he wrote about it for the U.S. Reclamation Service, notably in *Better Business, Better Farming, Better Living: Hints from a Practical Farmer,* published in 1918.

After financial setbacks in the 1920s and 1930s, O'Donnell's grandsons began building Hesper Farm back up. During the 1940s, O'Donnell Bros. produced wheat, Great Northern beans, sugar beets, oats, and alfalfa and started raising turkeys. The turkeys soon became a major part of their business, and a large, one-story turkey shed was added to the side of the barn.

An early 1900s photo of the Hesper Farm shows loads of sugar beets heading to the beet dump at Yegen in 1906, the year the Billings Sugar Company started operating.

Harley O'Donnell (above) stands by the door of the old bunkhouse, where he maintains a library (below).

Weldon Barn

YELLOWSTONE COUNTY, LATE 1920s

From territorial days well into the 1960s, most Montanans got their milk from a local source. Every city had a few large dairies, but there were also lots of farmers who kept a small herd and supplemented their income through milk sales. As Yellowstone County ranchers Orville and Elizabeth Weldon explained, "The dairy was the way we could afford to send our older children to high school."

Monica Weldon with some of the family livestock

Orville, Elizabeth, and children George, Ben, Charles, Clinton, Maeona, Delores, and Dick all ran the dairy as a family. In the wee hours each morning before school, the boys fed and milked the cows by hand and delivered the milk to houses and businesses in Billings. The boys fed and milked again in the evening while the younger children helped Elizabeth to bottle milk and sterilize equipment. All of this work was done without the convenience of running water (there was a well in the farmyard) and without electricity. Kerosene lamps lit the barn until Rural Electrification Association power lines reached the farm in 1946.

The Weldon Barn was built in the late 1920s with recycled materials from a lumberyard in Billings and from an old barn on the place. Throughout the 1930s and 1940s, the barn housed the Weldon milk cows and eight large workhorses. The youngest Weldon, Dick, eventually took over the ranch and operated a dairy into the 1950s. Today, the barn still gives shelter to mother cows and calves during harsh winter weather. It also houses chickens, goats, a pony, burros, and donkeys, along with the family's saddles, bridles, and ropes.

This little barn, so humble and simple in its construction and function, embodies not just the Weldon family's story but also symbolizes the culture of work and family that became the legacy of small farms and ranches across the vast rural landscape of Montana.

A
W

Miller Ranch Barn

STILLWATER COUNTY, 1890

The Miller Ranch Barn, with its distinctive textured shingles on the roof and gable ends, is almost all that is left of the Miller Ranch and the pioneer-era stockyards at Merrill, Montana. The barn was built in 1890 for Matthew and Elsie Miller, Swedish immigrants who came first to Illinois and then, in 1881, to Montana. Elsie Miller's uncle, Olof Lafverson, was an early pioneer who settled at Merrill in 1878, then just a stop along the Bozeman–Miles City stagecoach line northwest of present-day Columbus.

When the railroad came through in 1882, one of the biggest stockyards along the line, that of the Montana Cattle Company, built its first rail outlet at Merrill. The Miller Ranch became the first post office and the center of social activity for the valley between Columbus and Big Timber. According to *They Gazed upon the Beartooths*, the Millers' home had to be enlarged to entertain the guests. The barn housed draft horses and saddle horses and, during shipping season at the stockyards, guests and crew as well. A sheepshearing barn, a dairy barn, and a smaller version of the horse barn are now gone.

After Matthew and Elsie Miller died in 1909, their son, John, and the younger Miller children stayed on the ranch. The historic photo of the ranch, taken in early spring 1919, captures the Millers' large two-story, wood-frame house, the centerpiece of the ranch. The barns, though equally as impressive, are set at a respectful distance from the house and are screened by trees to separate the domestic and farm operations. Mules, horse teams, and hounds are all on display in the picture, and the Miller children and their friends look quite smart on the tennis court (far left), a rare sight in early-day Montana.

This 1919 view of the Miller Ranch is a classic portrait of a prosperous Yellowstone Valley ranch, from the tennis court on the left to the matching barns in the background on the right.

Blacksmith shop interior

OW Ranch Barn

BIG HORN COUNTY, CIRCA 1890

This unusually large multicrib, one-story barn stands out among the vernacular log buildings of eastern Monana. Built in the late 1880s, it gives evidence of fine craftsmanship in the hand-hewn logs and tightly notched corners. In addition to the precise log work, the one-story design with connecting corrals and workspaces is ingenious and efficient. Instead of many separate buildings and corrals, this U-shaped barn incorporated separate holding pens and a divided courtyard corral for horses and cows, with enough space for a

Little has changed structurally at the OW Ranch since this photo was taken in 1895.

smokehouse and a blacksmith shop—all under one roof. This efficiency of design is present in all of the log and stone buildings on the property, perhaps reflecting one of the reasons this ranch is among the most successful in southeastern Montana.

Rancher John Kendrick first managed the OW in 1889 for the Converse Cattle Company, supervising construction of all the buildings and overall ranch operations. He went on to buy the ranch in 1898, launching a cattle empire in Montana and Wyoming. By 1917, he was running over fifteen thousand head of cattle, though he sold most of them in 1919, foreseeing the end of a favorable market after World War I. Kendrick was elected governor of Wyoming in 1914 and became the state's U.S. senator in 1916. He served as a senator until his death in 1933.

Photographer L. A. Huffman visited the East Fork of Hanging Woman Creek and captured the ranch in its heyday in August 1890, at a time when, as historians Joan Brownell and Kirk Michels wrote in the ranch's National Register of Historic Places nomination: "There was always a friendly greeting, a dry place for your horse, and an extra plate at the table. . . . Today, the OW's pristine condition documents the day-to-day life on a working cattle ranch, where the cook still rings the bell for dinner and the cowboys still sleep in the bunk house and ride their horses up the draw to check on the range."

(center top) The left side of the U-shaped building housed the horse barn, the right side was the blacksmith shop, and the connecting section was used for calving.

The Kukowskis, posed here on horseback, stable their trusty horses in the old N-Bar Ranch Barn.

Faddis-Kennedy/N-Bar Ranch Barn

BIG HORN COUNTY, 1939

Organized in 1913, the Spear-Zimmerman Cattle Company was a major presence in the Lodge Grass area and on the Crow Reservation for many years. The ranch acquired substantial ranges in Montana and Wyoming, and in 1927 the company reorganized as the Faddis-Kennedy Ranch. A portion of the famous N-Bar Ranch, one of the largest spreads in the West in the late nineteenth century, was added to the ranch, and a shadow of that history remains today in the fading N-Bar brand displayed on the Faddis-Kennedy Barn.

This handsome barn was built in 1939 by Edgar R. Fenton with lumber from the Johnston Lumber Company in Lodge Grass. Fenton, who was blind in one eye, built the barn by himself, cutting all the lumber and working through most of the winter to finish it. The horizontal-lapped siding, white corner boards, gabled ventilators, and evenly spaced square windows suggest that the barn design may have come from a commercial pattern book or a barn-plan service. Half of the barn was used for dairy cows; the other half for horses, harnesses, saddles, and tack essential to ranching operations.

Rugg Barn

BIG HORN COUNTY, CIRCA 1915

Like many homesteaders, when the Rugg family came to southeast Montana in the early 1900s, the first thing they built was a log barn. Several years later, when the ranch was more established, they built this sturdy stone-and-wood-frame barn, marking it proudly with the family's registered brand. The J-O brand was first recorded in Wyoming Territory in 1888 and later recorded in Montana in 1920 by the Rugg family.

The barn is a well-kept landmark in the Kirby area and is especially noticeable for its cow weather vane—an appropriate topper since livestock and ranching are a prominent and important part of Kirby's past and present.

Three Circle Barn

ROSEBUD COUNTY, 1916

In 1884, Ed Brown, a Mississippi native, founded a ranch on the Tongue River near Hanging Woman Creek; two years later, his brother, Captain Joseph T. Brown, a veteran of the Confederate Army, trailed in the first herd of cattle, one thousand head, from what was then Indian Territory (Oklahoma) to establish the Three Circle Ranch.

Joseph Brown purchased the Zook-Alderson homestead a couple of miles up the river and moved the Three Circle home ranch headquarters there. In 1902, Joseph and his wife, Mary, hired masons from Sheridan to build a substantial stone ranch house, bunkhouse, and outbuildings. In 1916, they added a massive stone horse barn, a landmark in the region and one of the most impressive barns in the state.

The main barn had room for draft horses, with open stalls off the rear door that led to a loafing shed and a stone chicken enclosure. A wing to the east provided space for wagon storage and a blacksmith shop. The barn was banked, and the enormous loft could be accessed from either an upper side entrance or through hay doors on the front. To crown this impressive building, stonemasons mounted matching buffalo skulls on the front walls. Art Hayes Jr., great-grandson of Joseph T. Brown, still runs the ranch with help from his wife, Marilyn, and his sons.

Art Hayes Jr.

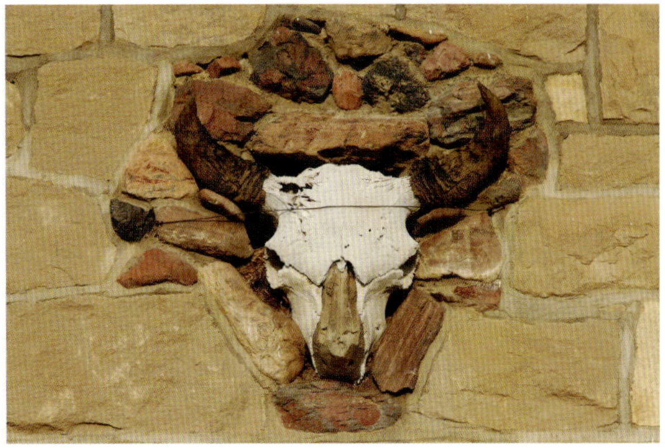

Pictured here (left to right) are Irving Alderson Sr. ("Little Bones"), Allen Roberts Alderson ("Big Bones"), Percy Roberts Cox, and Edwin Allen Cox.

Bones Brothers Ranch Barn

ROSEBUD COUNTY, 1924

The large sandstone-and-shingle-sided barn on the Bones Brothers Ranch was a keystone in the operation of this well-known cattle and dude ranch in the Hanging Woman Valley. On the roof, the ZC Bones TH brand is clearly visible, a reminder of the early days when Zachary Cox bought 320 acres of prime ranch land along the Hanging Woman Creek. The name "Bones" was a nickname given to Floyd Alderson along with his little brothers, Allen "Big Bones" and Irv "Little Bones," when they were teenage horse wranglers with the Circle Three wagon in the early 1900s.

The Bones brothers opened their ranch to guests in 1924 after spending three years working in Wyoming at the Eaton Ranch, one of the first dude ranches in the West. Cattle ranching took a downturn after the harsh winter of 1917, which was followed by years of drought that left many ranchers and ranch hands looking for new ways to make ends meet.

This sandstone barn was built to launch the dude ranch operation in 1924, along with twenty-four other buildings, including guest cabins and a rec hall to accommodate out-of-town guests. From the barn, the dudes would saddle up and ride out to help with ranch chores or go on trail rides. Over the years, the sandstone walls served as a diary of sorts for cowhands and dudes who scratched their names into the soft stone walls, leaving their mark while waiting for their horses and the day's adventures. While the ranch has not taken guests since the 1960s, the barn remains in daily use, part of the cattle operations still carried on by descendents of the Bones brothers, Irv Alderson and his family.

Murphy/Mizpah Ranch Barn

POWDER RIVER COUNTY, 1941

The Mizpah Ranch history traces the ups and downs of Montana agriculture, from the promising open-range days and the homestead boom, through plunging years of drought and the Great Depression, and back up with the rebounding economy during and after World War II. The Mizpah Barn was built in 1941 during this hopeful period as the ranch became one of the largest in eastern Montana.

Beginning in the 1880s, the Powder River country was controlled by a few large livestock companies that ran cattle across hundreds of thousands of acres of what was then Custer County. John M. Holt ran the Mizpah Livestock Company, which controlled rangelands in the Mizpah, Powder, Upper Box Elder, and Cabin Creek drainages. Backed by Missouri investors, Holt and his partner, John T. Murphy, acted early on to acquire many smaller ranches and buy vast sections of land from the railroad as soon as it became available, consolidating a large private range for cattle and sheep raising. But with the drought and depression of the late 1910s and 1920s, Holt and Murphy's livestock empire faltered, and in 1931 they lost their land to the Federal Land Bank.

In 1941, New Mexico cattleman John McNierney bought the former Mizpah lands, about 64,000 acres, from the government. A few months later, McNierney made headlines when he purchased the Clyde Brown Ranch, consisting of over 128,000 acres on the Powder River, in one of the largest sales of private land in the state's history.

That same year, McNierney hired men to put up a classic red barn and house for his ranch foreman. The gambrel-roof barn is clearly a twentieth-century structure, with a focus on efficiency evident in its design. The interior has a cross plan, with a wide entrance on the eave side and a narrow alley down the center that is lined with twelve stalls for horses; tack and grain storage rooms stand near the center. Chutes in the hayloft allowed hay to be dropped through openings in the floor directly to the horses below. Hay loading was quick and easy given that the rear side of the barn was on a bank, allowing a truck or tractor loaded with hay to pull up directly to the loft doors and unload right into the expansive hayloft.

Nunberg Barn

WIBAUX COUNTY, 1911

The Nunberg Barn, built in 1911, is a classic red gambrel-roof barn that has housed a dairy operation, workhorses, feed, and community social functions. First purchased by Alfred Nunberg Sr., the property the barn stands on has been in the Nunberg family since 1928 and is still owned by Shirley and Fred Nunberg. Born on the farm in 1929, Fred (Alfred Jr.) eventually took over from his parents after World War II and raised wheat and livestock until the 1990s.

The Nunbergs were Polish homesteaders, part of a community of Poles who were active in St. Phillip's parish in Wibaux. Though the Nunbergs bought this property at the height of the Great Depression, they managed to weather those hard years and expand the farm. Fred's father, Alfred Sr., ran a diversified operation, raising pigs, chickens, sheep, beef and dairy cattle, wheat, barley, oats, and hay. Fred remembers that the train stopped every day in Wibaux to pick up the cream cans that were then shipped to the creamery at Mandan, North Dakota.

From 1934 to 1938, the family hosted community barn dances on a regular basis to supplement the farm income. One of the bands that played at these dances featured William Burns Jr., who later went on to play with well-known "big bands" like those of Tommy Dorsey and Benny Goodman. During the early 1940s, with the drought and grasshoppers plagues in the past, Alfred Sr. enjoyed good years for grain growing, and much of the grain was stored in the barn in the space vacated by the horses, which had been traded for a tractor.

The Nunbergs received a grant from the Montana Historical Society's State Historic Preservation Office in 2006 to repair the barn's foundation, siding, doors, and windows. With these funds, the Nunbergs were able to carry on the legacy of their forebearers and keep their barn in use.

Tin plates by the hundreds join the butt-ends of the silo siding planks.

Detail of a rafter bent in the haymow and the pegged mortise-and-tenon joint

Wicka Barns

WIBAUX COUNTY, 1895, 1914

These two Wibaux County barns represent two very different eras of agricultural advancement, and both are outstanding barns of their type in the area.

The classic gable-roof horse and cattle barn is an excellent example of a gable-entry barn with timber framing. It was likely an early farm building in the area, built in the late 1890s or early 1900s when settlers from the midwestern states and parts of the Dakotas were moving farther west to newly opened homestead lands. The barn has many characteristics of the Old World English threshing barn but was adapted to house draft horses and cattle. Instead of having a large opening on the eave side of the barn and a wide central alley for threshing, it has two small offset doors at one end under the gable, a large hay door on the east, and a smaller hay door on the south. Inside, narrow side alleys provided more space for stalls along the sunny and warm south wall and down the center of the barn. Upstairs in the hayloft, the barn's heavy-timber bracing clearly shows expert craftsmanship in the uniformity of the square posts and in the perfectly matched and pegged mortise-and-tenon joints. Remarkably, some of the pencil marks left by the carpenter are still visible over one hundred years later.

In contrast to the Old World character of the horse and cattle barn, the hexagonal silo barn, built in 1914, represented the latest in animal-feeding technology at the time. On the exterior of the barn, a metal plate indicates that the barn was designed and possibly built by the Independent Silo

(above) The classic gable-roofed barn was built in the late 1890s or early 1900s.

(left) The hexagonal barn was built in 1914.

(far left) The all-wood circular silo rises from the ground and continues through the roof of the hexagonal barn. The silo's conical roof has collapsed into its interior.

Company; patent dates of 1903, 1913, and 1914 are found on the plate. The all-wood circular silo was invented by University of Wisconsin professor of agricultural physics Franklin King in the late 1880s. Professor King designed a round barn for his brother that was built of horizontal wood siding with an internally supported conical roof and an interior circular silo. King became a leading proponent of round barns, and his silo design was widely accepted and promoted as the "King of Wisconsin Silos." This system was touted as the most efficient, healthful, and sanitary method of cattle feeding, but the up-front costs of building proved to be too expensive for many Montana farmers and ranchers.

As the round-silo barn concept gained widespread popularity, companies such as Independent Silo of St. Paul, Minnesota, marketed their own versions of the King silo barn. Generally, round-silo barns were used for dairy cows; however, the Wicka silo barn is bare inside, with no sign of stalls or stanchions, indicating that it was used only for occasional shelter and feeding. Locals report that there used to be one or two more barns like the Wicka silo barn in the area, but this is the only one of its kind left in Montana.

Johnson Horse Ranch Barn

PRAIRIE COUNTY, 1904

Although today the Johnson Barn is losing its battle against the prairie winds of eastern Montana, barns like this were once symbols of opportunity for those seeking a prosperous farming life on "free" lands in the West. J. D. Johnson brought his family from Illinois to the Spring Creek area east of Fallon in 1896. The Johnsons started out with a small shack and a crude log barn; they raised cattle and horses and belonged to the roundup pool. After a few years of good rain and a lot of hard work, Johnson hired other homesteaders, the Hogmires and Omsteads, to build him a handsome barn.

The gigantic barn is noteworthy not only for its size but also for its distinctive pressed-tin siding. At the turn of the nineteenth century, pressed-tin siding became fashionable, mostly in residential settings, as a durable and attractive covering for houses and garages. However, Johnson likely chose the pressed tin not for fashion but because it was more likely to stand up to the harsh sun, wind, and freezing temperatures of the prairie.

In 1913, eastern Montana's famed photographer Evelyn Cameron took several pictures of the Johnson Ranch when it was looking its best. The Johnsons leased the ranch in 1917 or 1918 and then sold it in 1925 when they retired to California. After more than one hundred years on the prairie, the barn still stands up to the test of time.

The only people identified in this Evelyn Cameron photograph, taken at the Johnson barn in 1913, are J. D. Johnson on horseback, Mrs. J. D. Johnson on the far right, and Mrs. Johnson's daughter, Mrs. Tom Renn, on the far left. The other woman is not identified.

Detail of pressed-tin siding

Pehl Barns

PRAIRIE COUNTY, 1912, 1918

William and Andrew Pehl were among many farmers in the Midwest and Dakotas who pushed west into Montana in the early 1900s. The brothers came from Ellendale, North Dakota, to Terry in 1909, purchasing 1,280 acres of Northern Pacific Railroad land for fifteen dollars an acre. Several years later, William Pehl testified in favor of the railroad in federal court in Washington, D.C., in a lawsuit that charged the railroad with making false claims about the quality of the land they sold to homesteaders.

After purchasing land, the Pehl brothers summered in Montana and wintered in North Dakota—plowing, planting wheat, and making improvements to each of their properties. William Pehl hauled his steam-engine tractor from North Dakota to use in the fields. But, instead of hauling all the lumber for a house out to the farm, he first built the house on skids on a lot in Terry, then pulled it to the farm with the steam engine.

By 1914, the brothers were living on their farms full time. Andrew Pehl built his barn first, around 1912, the first one on this part of the Broadview Bench. The first floor held draft horses under the steeply pitched saltbox-style roof. The relatively small hayloft held just enough feed for the horses. Pehl farmed here until the late 1920s before selling to Frank Haegle, who, with his family, operated it until 1991.

Brother William Pehl's barn was finished in 1918, as the twin ventilators with their horse weathervanes attest. This classic gambrel-roof barn held draft horses along the center

alley and milk cows in the stalls on the side. William, like others in the area, raised spring wheat, but he was best known for his hog operation. In November 1924, the *Billings Gazette* lauded William Pehl as one of Prairie County's "most successful hog raisers," describing how Pehl had moved his hogs the six miles from his farm to the Terry depot: "Mr. Pehl says that his hogs walked to Terry easier and cheaper than he could have loaded and trucked them in, and arrived in as good condition."

After the farm was handed down to Pehl's son John in the mid-1950s, the barn was used for grain storage. Today, both barns are owned by William's grandson. They stand as symbols of the boom days of wheat farming in eastern Montana.

(opposite page) Andrew Pehl's 1914 barn is sided with cedar.

(this page) William Pehl's barn, 1918

(left) Chicken barn

Malkuch Dairy Barn

DAWSON COUNTY, 1940s

Wisconsinites Frank and Esther Malkuch came to Montana in 1924 and began farming. After many years of hard work, they built this large dairy barn west of Glendive in the 1940s. Since Esther came from a long line of dairy farmers, she wanted a dairy barn like those back home.

Indeed, this dairy barn, in technology and design, took on many of the qualities of the large dairies in the Midwest and is a good example of a "modernized" early-twentieth-century dairy barn. It had space for a large herd and was built using the latest materials of the day—steel posts and stanchions, steel waste grates, and prefabricated, laminated and bent rafter beams in a center-aisle layout prescribed by many Extension programs and barn-plan services. The story goes that Esther did all the milking because Frank, who was mostly deaf, would shout in the barn and scare all the cows.

Two railcars form part of the structure of this barn, one across each end. To access the center of the barn, one passes through doors on either side of the railcars. This view is photographed from the doorway area toward the end, showing the railcar's roof structure. One railcar space is used for storage and the other for tools and a workshop.

Oljar/Klempel Barn

DAWSON COUNTY, 1934

The Oljar/Klempel Barn north of Glendive is a recycled barn, built from the remains of the Yates grain elevator, which was torn down when Yates fell victim to hard times and disappeared from maps during the Great Depression. Established in 1888, Yates was located 4.9 miles east of Wibaux. As the first railroad stop in Montana, Yates became a bustling town in the early 1900s and boasted two grain elevators. Drought struck in the late teens, and by 1930 farm failures had forced many business closures. The buildings at Yates were moved or torn down for their lumber.

In 1934, John Horst, an early merchant and stock raiser in the Glendive area, transported two boxcars carrying the lumber, nails, and other remains of one of the Yates elevators to Glendive and recycled them into a barn. The boxcars make up the ends of the barn, and the other materials went into construction of a sturdy gambrel-roofed building. In 1946, John Oljar bought the barn to use for livestock operations in connection with his slaughterhouse. The barn is still used today and is well loved by its owners, Don and Debbie Klempel, who re-sided it with shiplap in 2000 and put on a new metal roof in 2001.

Diamond D Barn

RICHLAND COUNTY, CIRCA 1911

Christ and Carl Johnson, sons of Norwegian immigrants to Minnesota, ventured out west as young men. While in North Dakota, Carl died, and Christ decided to keep moving. In 1905, on his twenty-first birthday, he settled down in Montana. With the little money he had in hand, he filed on a 160-acre homestead five miles northwest of Fairview.

Just putting up a fence around the property was a major chore, but it was a strong start to his new farm. Johnson built a small house in 1906 and began farming in earnest in 1908. The railroad came through Fairview in 1911, and the barn and other outbuildings were likely built in the mid-1910s when good rainfall and a bright economy resulted in much building in Richland County.

The Johnson farmstead represents a successful Mon-Dak area farm, with its good-sized horse barn, windmill, chicken house, machine shed/shop, and tractor barn, all painted red and placed in close proximity to a two-story farmhouse. Like all farmers, the Johnson family had bountiful years and bad ones. In 1928, the haymow overflowed with sweet clover that had grown four feet high, while the next decade brought horrific conditions. At times, the entire wheat crop was wiped out by grasshoppers or hailstorms, and for many years, Christ and his wife, Bessie, worked off the farm to make ends meet.

The Johnsons' story is common across Montana and many states in the West. Farming was backbreaking, unpredictable, and never ended at dinnertime, but those who persevered often found their reward in working closely with their families and in learning the secrets of prosperous farming on the northern plains.

(above) The next generation of ranchers

(top left) The boxy structures against the loft wall are hay chutes for dropping hay to animals in the barn below.

Doyle Barn

RICHLAND COUNTY, 1906

Ben Doyle homesteaded here in 1903 and hired Joe Belanger in 1906 to build this bank barn to shelter his Percheron horses. The lumber and materials cost $868 back then, or about $20,000 today; the story has been handed down that the lumber estimate was so close to the actual use that there was not even a bushel basket of scrap left over.

The Doyle Barn was built against a hillside so that horses could pull a wagon into the huge hayloft. After the hay slings were elevated above the wagons, they were pulled to the front or back of the loft and released, allowing Doyle to save labor and stack the hay to the rafters. While bank barns are less than common in Montana, quite a few were built near the North Dakota line as relocated farmers from the Midwest brought their barn designs with them.

The barn has stood up to years of abuse by nature out on these wide-open plains. In 1947, a tornado blew through, turning the barn on its foundation, but contractor E. W. Chase was able to save the barn by coaxing it back to its original alignment with the use of cables and tractors. Today, the Wheeler family still uses the barn for calving and hay storage and proudly maintains it as the second century of its history unfolds.

The barn got a new coat of paint in 1941, applied by the Borg brothers, Menter and Gilbert.

CHAPTER 7

Preserving the Legacy

Any jackass can kick down a barn, but it takes a good carpenter to build one. —SAM RAYBURN

"That old barn on your place, Charlie,
Was nearly falling last time I saw it.
How is it now?"
"I got some poles to hold it on the east side
and the wind holds it up on the west."
—from *The People Yes,* CARL SANDBURG

THE BARNS of Montana are teeming with stories, reflecting the ways in which, across our history, people forged a partnership with the land. Barns tell of past decades when horses pulled plows that turned the sod, oxen and mule teams hauled freight, and mutton was on the dinner table of most Americans on a regular basis. They also speak to generations of hard work, as people balanced their lives against the forces of nature, the bounty of the earth, and whatever luck came their way.

The people who achieved this balance survived and stayed on the land. Those who could not sustain the equilibrium generally sold out and moved on. Whether they stayed or left, their experiences are a part of the story of Montana and of its people.

Today, the need for teams of large workhorses and scores of cowboys has faded into the past, and the use of cavernous storage buildings has diminished. To preserve a barn in this day and age is both an act of pride and a financial commitment. The sheer size of a barn and nature's constant weathering can make painting, roofing, and general upkeep a challenge.

Still, Montana is a place where horses and cows and chickens are a part of life and of the culture. Preserving a barn through continued use is not just a possibility, it is an occupation for many Montanans. With over six thousand historic barns still standing, we have the chance to make the most of the state's agricultural legacy.

There are many ways to preserve a barn, whether by keeping it active in ranch and farm operations, by adapting it to a new use, or by moving it to new ground. In fact, relocating barns and other rural buildings has been common for the last

(opposite) Ventilator and hawk decorate a barn in Gallatin County.

What could be reused was never wasted on the Montana prairie. A moving crew relocates a barn in Browning using a gas tractor circa 1912.

The historic Slack Barn (above) on the Teller Wildlife Refuge near Corvallis is now an active interpretive center used for workshops, volunteer projects, and, on many summer weekends, weddings.

(above left) The zip line is the latest in hayloft recreation.

150 years. With a pragmatism born of necessity, Montana's farmers and ranchers never let much go to waste, especially useful, well-constructed buildings. Moving sheds, barns, cabins, and schoolhouses by horse team, steam tractor, or semi-trailer to new locations to repurpose them has long been part of life in Montana.

Sensitively converting these rural buildings to new uses preserves a heritage and offers a new chance for many carriage houses and barns to serve as homes, shops, offices, reception halls, and museums. This kind of adaptation has saved buildings like the Slack Barn, the Springhill Pavilion, and the Lewistown Art Center. It is in the future for the Adams Stone Barn and many others.

Today, the hope for our barns lies in the hands of rural Montanans and the many friends who love old buildings and who work to preserve them. To that end, we hope that this book will inspire people and remind them that all such buildings, hand raised and well worn, reflect a timeless beauty that comes with age, hard work, and a quality of craftsmanship that is disappearing from today's world.

Sisters on the Jemison Ranch were proud to share their barn and a glimpse of their new foal with visitors on a Cascade County barn tour.

Notes

Chapter 1

1. In old Europe, barns and stables were often attached to the home or were built on the ground floor with the family living upstairs. However, over time, European ethnic groups arriving in North America generally adopted the English habit of moving animals into different buildings.
2. Jordan, Kilpinen, and Gritzner, *Mountain West,* 127–29.
3. See Vlach, *Barns,* 17. Calkins and Perkins also offer analysis of the English barn, sometimes referred to as the three-bay threshing barn. See their essay, "Three-bay Threshing Barn," 40–61.
4. Transverse-crib barns became in many minds the classic American barn. Vlach, *Barns,* 19. See Noble and Wilhelm, *Barns of the Midwest,* for a full discussion.
5. Jordan, Kilpinen, and Gritzner, *Mountain West,* 37–40.
6. Ibid., 38, 127–29.
7. Vogel, "Common Farm Barns," E-22. Like the sod barns that Vogel describes in neighboring South Dakota, Evanson's sod barn, pictured here in 1905–1907, was small and unfinished on the exterior, unlike many sod houses whose outside walls were mudded.
8. Sievert and Sievert, *Guide to Historic Lewistown,* 63–71, and National Register nominations on file with SHPO for the Helena South Central Neighborhood and the White Face Farm (Kleffner Ranch).
9. Hubka, "Americanization of the Barn," 2.
10. Radford, *Radford's Practical Barn Plans,* 64; Vogel, "Common Farm Barns," E-8. Vogel's work in neighboring South Dakota provides an excellent foundation for understanding national barn history and regional developments common to the High Plains states.
11. Radford, *Radford's Combined House and Barn Planbook,* 3. Vlach suggests that the broad appeal of Radford's books in part stemmed from the fact that the designs were familiar and embraced such formats as the three-bay barn, the bank barn, and the transverse-crib barn. Vlach, *Barns,* 9–28.
12. Vlach, *Barns,* 24.
13. Leffingwell, *American Barn,* 106–107.
14. Vogel, "Common Farm Barns," E-14, F-47-48; Soike, "Within the Reach," 158–59.
15. James Manufacturing, *James Way,* Catalog 22.

Chapter 2

1. Burlingame and Toole, "Range Livestock Industry," 311.
2. Ibid.; Barbour, *Fort Union,* 118, 125–26.
3. Dunbar and Phillips, eds., *Journals and Letters,* vol. 1, 281; Bell, Schaffer, and Richards, "1860 Census"; U.S. Department of Commerce, *Ninth Census: 1870.*

4. John Milner Associates, "Grant Kohrs Ranch," chap. 2, 18.
5. K. Ross Toole noted that there were factors beyond the Stuart brothers' discovery that led gold seekers to swarm into Montana. Toole, *Montana,* 65–69.
6. Burlingame and Toole, "Range Livestock Industry," 313, 317.
7. In 1892, the longest freight train on record—seventy-two oxen pulling ninety-six thousand pounds of wool—hauled its load to Big Timber. Malone, Roeder, and Lang, *Montana,* 77.
8. Burlingame and Toole, "Range Livestock Industry," 313; Briggs, *Frontiers of the Northwest,* 209.
9. Briggs, *Frontiers of the Northwest,* 209–11.
10. Malone, Roeder, and Lang, *Montana,* 155.
11. Briggs, "Development and Decline," 527.
12. Malone, Roeder, and Lang, *Montana,* 156–57; Briggs, *Frontiers of the Northwest,* 216–18.
13. Briggs, *Frontiers of the Northwest,* 319.
14. Malone, Roeder, and Lang, *Montana,* 166; Burlingame and Toole, "Range Livestock Industry," 325.
15. Burlingame and Toole, "Range Livestock Industry," 327; Malone, Roeder, and Lang, *Montana,* 167.
16. The National Register nomination on file with SHPO for the Ninemile Remount Depot offers much background on this site.
17. Campbell, *Campbell's 1907 Soil Culture Manual.*
18. Professor Maire O'Neill and students at Montana State University's School of Architecture studied numerous barns in Gallatin County. At the turn of the twenty-first century, of the barns still standing, wood-frame barns outnumbered early log and timber-framed barns by roughly four to one. O'Neill, "Inventory of Historic Agricultural Structures," 69–70.
19. Malone, Roeder, and Lang, *Montana,* 283.
20. Libecap explores this epic of hardship in detail in his article "Learning about the Weather," 24–33.

Chapter 3

1. Bancroft, *History of Washington, Idaho, and Montana*, 603.
2. Ibid.
3. Bell, Schaffer, and Richards, "1860 Census"; Dunbar and Phillips, eds., *Journals and Letters,* vol. 1, 216.
4. Lewis's journal entry of July 3, 1806, in Moulton, ed., *Journals,* vol. 8, 86.
5. Meagher, "Rides through Montana," 583; Frenchtown Historical Society, *Frenchtown Valley Footprints,* http://www.usbr.gov/dataweb/html/frenchto.html#N_7_.
6. Chittenden and Richardson, eds., *Life, Letters & Travels,* vol. 4, 1233, 1241.
7. Burlingame and Toole, "Range Livestock Industry," 311.

Chapter 4

1. Stauffer, ed., *History of Beaverhead County,* 118–19; Montana Department of Environmental Quality, *Historical Context.*
2. Heath, "Striving for Permanence," xi–xxxiv. In the preface to his important study of the evolution of false-front buildings, Heath discusses the meaning of vernacular or "everyday" buildings and how such functionally minded buildings reveal much about the cultural sense that guided builders on the frontiers of settlement. It is interesting to note that false-front barns were not common in Montana, and those that were built were often liveries.
3. The territorial capital moved from Bannack to Virginia City in 1865, providing a measure of stability to the boomtown. In its heyday, there were 1,200 buildings; today, 237 major buildings remain. Among the cabins, saloons, shops, and gambling houses were stone, log, and frame barns to house the stock. Among them, the Kiskadden Barn and the Virginia City Livery stand out along the main street, their false fronts blending with neighboring buildings. Briggs, *Frontiers of the Northwest,* 20.
4. Burlingame and Toole, "Range Livestock Industry," 312.
5. *Dillon City . . . Directory*, 86; Davison and Myers, "Terminus Town," 24–26.
6. *Dillon City . . . Directory,* 86.
7. Burlingame and Toole, "Range Livestock Industry," 334–38, and Briggs, *Frontiers of the Northwest*, 338–45, both offer background on the origins of the state's equine industry.
8. Beaverhead County Museum Association, *More History*, vol. 2, 44–45; Malone, Roeder, and Lang, *Montana*, 150–51; U.S. Department of the Census, *Census of Agriculture,* 1900, 1910, 1920, 1930.
9. Two good sources about beaverslides are Jordan, Kilpinen, and Gritzner, *Mountain West,* 105–116, and Alwin, "Montana's Beaverslide," 42–50.
10. John Milner Associates, "Grant Kohrs Ranch National Historic Site, Cultural Landscape Report," website.

11. Harkness, "Diary," 353
12. Fisk, "Expedition," 25.
13. Jenks, *Guide to Historic Bozeman*, 17
14. Ibid.
15. Kroes, *Persistence of Ethnicity*, 50.
16. Knudsen, "Barns as Index," 40–48.
17. Murdock, *Farm Buildings*, 2.

Chapter 5

1. Lewis's journal entry of July 11, 1806, in Moulton, ed., *Journals*, vol. 8, 106.
2. Originally called Fort Lewis, the site was renamed Fort Benton in 1851.
3. Barbour, *Fort Union*, 116, 118, 125–26.
4. Bowen, *Progressive Men*, 145; Renewable Technologies, "Vaughn-Sun River Highway Project," 9.
5. As quoted in Brownell, "Cultural Resource Survey," 20.
6. Malone, Roeder, and Lang, *Montana*, 153.
7. Messiter, *Sports and Adventure*, 287–88, 346.
8. Malone, Roeder, and Lang, *Montana*, 167.
9. Gunderson, "High Line," 95.
10. Stuart, *Forty Years*, 22.
11. Worthen and Mueller, "Central Montana," 154; Burlingame and Toole, "Range Livestock Industry," 318.
12. Fort Belknap Reservation Superintendent William R. Logan, 1905, as quoted in Newby, "Reinventing Identity."
13. Gunderson, "High Line," 97.
14. U.S. Department of Commerce, *Thirteenth Census: 1910*, 1152.

Chapter 6

1. Journal entry for July 24, 1806, in Moulton, *Journals*, vol. 8, 219.
2. Fort Remon is better known in history as Lisa's Fort or Fort Manuel.
3. West, *Capitalism on the Frontier*, 33–34.
4. Ibid., 58–59.
5. Ibid., 78.
6. Historical Research Associates, "Patterns of the Past," 44.
7. Brisbin, *Beef Bonanza*, 90.
8. John Milner Associates, "Grant Kohrs Ranch," chap. 2, 127.
9. As quoted from "Autobiography," 114–15, in John Milner Associates, "Grant Kohrs Ranch," chap. 2, 132.
10. As quoted from Historical Research Associates, "Patterns of the Past," 46.
11. Warhank, "Fort Keogh," 279.
12. Burlingame and Toole, "Range Livestock Industry," 318.
13. Brown, *Plainsmen of the Yellowstone*, 381–82.
14. Jiusto, "Camp 4 National Register Nomination," sec. 8; Fitzgerald, *Every Farm a Factory*.
15. Bernstein, *Families That Take in Friends*, 53.
16. Some illustrious Americans were listed on the guest rolls at Montana's dude ranches, including cowboy writer Will James, Mary Roberts Reinhart, and Gary Cooper. As Cooper wrote later in *The Dude Rancher Magazine*, "The dude ranches have given many people an appreciation of the West, which, though it may not be as wild and wooly as it was, retains all of its vigorous beauty. And running a dude ranch is my idea of a great occupation. I've tried it once and may have another fling at it one of these days." Bernstein, *Families That Take in Friends*, 72–73.

Bibliography

All information provided the authors by the owners of the historic barns featured in this book, including National Register nominations, is on file with the Montana Preservation Alliance (MPA) in Helena.

Books and Articles

Alwin, John A. "Montana's Beaverslide Hay Stacker." *Journal of Cultural Geography* 3, no. 1 (1982): 42–50.

Arthur, Eric, and Dudley Witney. *The Barn: A Vanishing Landmark in North America.* New York: Graphic Society, 1972.

Bancroft, Hubert H. *The Works of Hubert Howe Bancroft,* vol. XXI. *History of Washington, Idaho, and Montana, 1845–1889.* San Francisco: History Company, 1890.

Barbour, Berton H. *Fort Union and the Upper Missouri Fur Trade.* Norman: University of Oklahoma Press, 2001.

Beaverhead County Museum Association. *More History of Beaverhead County, 1800–1997,* vol 2. Dillon, Mont.: Beaverhead County Museum Association, 1997.

Bell, Margery H., Katherine Schaffer, and Dennis Richards, indexers. "1860 Census Bitter Root Valley & Ponderay Mountains Area of Washington Territory now in the State of Montana." Unpublished manuscript, n.d., PAM 131, Montana Historical Society (hereafter MHS) Library, Helena.

Bernstein, Joel. *Families That Take in Friends.* Sheridan, Wyo.: Dude Ranchers Association, 1983.

Bitter Root Valley Historical Society. *Bitterroot Trails,* vols. 1, 2, 3. Stevensville, Mont.: Bitter Root Valley Historical Society, 1982, 1998.

Bowen, A.W. *Progressive Men of the State of Montana.* Chicago: A. W. Bowen, 1903.

Briggs, Harold E. "The Development and Decline of Open Range Ranching in the Northwest." *Mississippi Valley Historical Review* 20 (Mar. 1934): 521–36.

———. *Frontiers of the Northwest: A History of the Upper Missouri Valley.* New York: Appleton-Century, 1940.

Brisbin, James S. *The Beef Bonanza; or How To Get Rich on the Plains.* Philadelphia: Lippincott, 1890.

Brown, Mark H. *The Plainsmen of the Yellowstone.* Lincoln: University of Nebraska Press, 1969.

Brownell, Joan. "Cultural Resource Survey, Judith Gap—Glengarry 100 KV Transmission Line." On file Montana Preservation Alliance (hereafter MPA), Helena.

———. "National Register Nomination for Bones Brothers Ranch." On file MPA, Helena.

Bureau of Land Management. "Class III Cultural Resource Inventory of Selected Bureau of Land Management Parcels in Blaine, Fergus, Hill, Chouteau and Toole Counties, Montana." 2003. On file Montana State Historic Preservation Office (hereafter SHPO), Helena.

Burlingame, Merrill G., and K. Ross Toole. "The Range Livestock Industry." In *A History of Montana,* ed. Merrill G. Burlingame and

K. Ross Toole, vol. 1, chap. XIII, 311–39. New York: Lewis Historical Publishing Company, 1957.
———, eds. *A History of Montana,* 3 vols. New York: Lewis Historical Publishing Company, 1957.
Byorth, John. "Report on Andersen Family Oral Histories." Unpublished manuscript, 1997. MHS Library, Helena.
Calkins, Charles F., and Martin Perkins. "The Three-bay Threshing Barn." In *Barns of the Midwest,* ed. Allen G. Noble and Hubert G. H. Wilhelm, 40–61. Athens: Ohio University Press, 1995.
Campbell, Hardy Webster. *Campbell's 1907 Soil Culture Manual: A Complete Guide to Scientific Agriculture as Adapted to the Semi-arid Regions: The Proper Fitting of the Soil for the Conservation and Control of Moisture and the Development of Soil Fertility; How Moisture Moves in the Soil by Capillary Attraction, Percolation and Evaporation; The Relation of Water and Air to Plant Growth, and How This May Be Regulated by Cultivation.* Lincoln, Neb. H.W. Campbell, 1907.
Chittenden, Hiram M., and Alfred T. Richardson, eds. *Life, Letters & Travels of Father Pierre-Jean De Smet, S.J.*, 4 vols. New York: Francis P. Harper, 1905.
Coues, Elliott, ed. *Forty Years a Fur Trader on the Upper Missouri: The Personal Narrative of Charles Larpenteur, 1833–1872*, vol. 1. Minneapolis: Ross & Haines, 1962.
Crofutt, Chris, and Stephen A. Aaberg. "30 KM Northeast of Glendive Northeast Class III Cultural Resource Survey Results in Dawson County and Richland County, Montana." Aaberg Cultural Resource Consulting Report for Montana Department of Transportation, 2002. On file MPA, Helena.
Davison, Stanley R., and Rex Myers. "Terminus Town: The Founding of Dillon, 1880." *Montana The Magazine of Western History* 30, no. 4 (1980): 24–26.
De Smet, Pierre-Jean, S. J. *Oregon Missions and Travels over the Rocky Mountains, in 1845–46.* New York: Dunigan, 1847.
Dillon City and Beaverhead County Directory, 1916–1917. Chicago: Polk & Co., 1917.
Dunbar, Robert G. "Agriculture." In *A History of Montana,* ed. Merrill G. Burlingame and K. Ross Toole, vol. 1, 281–98. New York: Lewis Historical Publishing Company, 1957.
Dunbar, Seymour, and Paul C. Phillips, eds. *The Journals and Letters of Major John Owen, Pioneer of the Northwest, 1850–1871,* vol. 2. New York: Edward Eberstadt, 1927.
Emmons, David M. *Garden in the Grasslands*. Lincoln: University of Nebraska Press, 1971.
Endersby, Eric, Alexander Greenwood, and David Larkin. *Barn: The Art of a Working Building*. New York: Houghton Mifflin, 1992.
Ethnoscience, Inc. "Fort Belknap to Havre: A Cultural Resource Inventory along US Highway 2." Unpublished report for Montana Department of Transportation, 2003. On file SHPO, Helena.
Ferguson, J. A., and L. P. Benedict. *Eighth Report of the Bureau of Agriculture, Labor and Industry of the State of Montana for the Year Ending November 30, 1902*. Helena: Independent Publishing Company, 1902.
Fisk, James L. "Expedition of Captain Fisk to the Rocky Mountains: Letter from the Secretary of War, in Answer to a Resolution of the House of February 26." U.S. War Department, 1864.
Fortier, Samuel, assisted by A. P. Stover and J. S. Baker. *Irrigation in Montana.* U.S. Department of Agriculture, Office of Experiment Stations, Bulletin No. 172. Washington, D.C.: Government Printing Office, 1906.
Francaviglia, Richard V. "Western American Barns: Architectural Form and Climatic Considerations." *Yearbook of the Association of Pacific Coast Geographers* 34 (1972): 153–60.
Frenchtown Historical Society. *Frenchtown Valley Footprints*. Missoula, Mont.: Mountain Press Printing, 1976.
Goodearl, George P. *Poultry Houses*. Montana Agricultural College Experiment Station, Circular 129, May 1925.
Gunderson, Carl M. "The High Line." In *A History of Montana,* ed. Merrill G. Burlingame and K. Ross Toole, vol. 3, chap. XXV, 85–105. New York: Lewis Historical Publishing, 1957.
Halsted, Byron D. *Barns, Sheds and Outbuildings*. New York: Orange, Judd Co., 1881; reprint, Brattleboro, Vt.: Stephen Greene Press, 1977.
Haney, Chuck. *Big Sky Barns: Grand and Historic Barns of Montana*. Helena, Mont.: Riverbend Publishing, 2007.
Hankins, Caneta, Skelley Hankins, and Michael Thomas Gavin. *Barns of Tennessee.* Nashville: Tennessee Electric Cooperative Assoc., 2009.
Harkness, James. "Diary of James Harkness, of the Firm of LaBarge, Harkness, and Company." In *Contributions to the Historical Society of Montana,* vol. 2, 343–61. Reprint, Boston: J. S. Canner, 1966.
Hart, Arthur A. *Barns of the West: A Vanishing Legacy*. Boise, Idaho: Historic Idaho, 1996.

Heath, Kingston William. "Striving for Permanence on the Western Frontier: Vernacular Architecture as Cultural Informant in Southwestern Montana." Ph.D. dissertation, Brown University, 1985.

Historical Research Associates. "Cultural Resource Inventory of Montana Department of Highways Project F 78-2(5)27, Absarokee-Columbus Stillwater County, Montana." On file SHPO, Helena.

———. "Patterns of the Past: A Brief History of the Ashland Birney Area." Unpublished report, 1980. On file SHPO, Helena.

Hubka, Thomas C. "The Americanization of the Barn." *Blueprints Magazine*, Spring 1994, 2–7.

———. *Big House, Little House, Back House, Barn: The Connected Farm Buildings of New England.* Hanover, N.H.: University Press of New England, 1984.

Hudson, John C. "Frontier Housing in North Dakota." *North Dakota History* 42, no. 4 (1975): 4–15.

James Manufacturing Co. *The James Way.* Fort Atkinson, Wisc.: James Manufacturing Co., 1917.

———. *Jamesway Barn Book No. 94A.* Fort Atkinson, Wisc.: James Manufacturing Co., n.d.

Jenks, Jim. *A Guide to Historic Bozeman.* Montana Main Streets Series. Helena: MHS Press, 2007.

Jiusto, Chere. "Camp 4 National Register Nomination," 1995. On file SHPO, Helena.

———. *A Guide to Historic Hamilton.* Montana Main Streets Series. Helena: MHS Press, 2000.

Jordan, Terry G., Jon T. Kilpinen, and Charles F. Gritzner. *The Mountain West: Interpreting the Folk Landscape.* Baltimore, Md.: Johns Hopkins University Press, 1997.

Jorgensen, Henry. *The Way It Was: The Childhood of One Boy and One Montana Danish Community.* Helena, Mont: Henry Jorgensen. In MHS Library, Helena.

Kilpinen, Jon T. "The Mountain Horse Barn: A Case of Western Innovation." *Pioneer American Society Transactions* 17 (1994): 25–32.

Knudsen, Harold S. "Barns as an Index to Ethnic Origins in Western Montana." Master's thesis, University of Montana, Missoula, 1969.

Kroes, Rob. *The Persistence of Ethnicity: Dutch Calvinist Pioneers in Amsterdam, Montana.* Urbana: University of Illinois Press, 1992.

Leavengood, David. "The Mountain Architecture of Robert C. Reamer." *Mountain Gazette* 47 (1976): 6–11.

Leffingwell, Randy. *The American Barn.* Osceola, Wisc.: Motorbooks International, 1997.

Libecap, Gary. "Learning about the Weather." *Montana The Magazine of Western History* 52, no. 1 (2002): 24–33.

Lucey, Donna M. *Photographing Montana, 1894–1928.* New York: Alfred A. Knopf, 1990.

Malone, Michael P., Richard B. Roeder, and William L. Lang. *Montana: A History of Two Centuries,* rev. ed. Seattle: University of Washington Press, 1991.

Malouf, Carling, and William Smurr. "Fort Owen—A Chronology." Unpublished manuscript, Montana State University, Bozeman, n.d.

Marshall, Howard Wight. *Barns of Missouri: Storehouses of History.* Jefferson City: Rural Missouri Electric Cooperative, 2003.

Mathews, Allan James. *A Guide to Historic Missoula.* Montana Main Street Series. Helena: MHS Press, 2008.

McCormick, Mary. "Archaeological and Historic Resources of Sheridan County, Montana." National Register Multiple Properties Nomination, 1993. On file SHPO, Helena.

McDonnell, Anne. "The Catholic Indian Missions in Montana." In *A History of Montana,* ed. Merrill G. Burlingame and K. Ross Toole, vol. 1, 97–109. New York: Lewis Historical Publishing Co., 1957.

Meagher, Thomas F. "Rides through Montana." *Harpers Monthly,* 1867.

Messiter, Charles Alston. *Sports and Adventure among the North-American Indians.* London: R. H. Porter, 1890.

Miller, R. F. *Care and Management of Sheep on the Farm.* Bozeman: Montana Agricultural College Experiment Station, Circular 31, November 1913.

Mills, James H., and A. C. Schneider. *First Report of the Bureau of Agriculture, Labor and Industry of Montana for the Year Ending November 30, 1893.* Helena, Mont.: Independent Publishing Company, 1893.

———. *Thirteenth Report of the Bureau of Agriculture, Labor and Industry of the State of Montana for the Years 1911 and 1912.* Helena, Mont.: Independent Publishing Company, 1913.

Montana Agricultural Experiment Station. *50 Years of Research, Huntley Branch Experiment Station.* Circular 230. Bozeman: Montana State College, July 1960.

Montana Preservation Alliance. "Cultural Landscapes of the Tongue River Valley, in Rosebud County, Montana." Unpublished report, 2007, on file MPA, Helena.

Moulton, Gary E., ed. *The Journals of the Lewis and Clark Expedition*, 8 vols. Lincoln: University of Nebraska Press, 1993.
Murdock, H. E. *Farm Buildings.* Montana Agricultural College Experiment Station, Circular 131, January 1926.
———, R. M. Merrill, and J. O. Tretsven. *Dairy Barns for Montana Farms.* Montana Agricultural College Experiment Station, Circular 130, June 1925.
Myers, Rex C. "Homestead on the Range: The Emergence of Community in Eastern Montana, 1900–1925." *Great Plains Quarterly* 10 (Fall 1990): 218–27.
Niven, F. "The Manhattan Malting Company—Developer of the Gallatin Valley." Unpublished paper given at Montana History Conference, Bozeman, 1989.
Noble, Allen G., ed. *Wood, Brick & Stone: The North American Settlement Landscape*, vol. 2. *Barns and Farm Structures.* Amherst: University of Massachusetts Press, 1984.
———, and Richard K. Cleek. *The Old Barn Book: A Field Guide to North American Barns and Other Farm Structures.* New Brunswick, N.J.: Rutgers University Press, 1995.
———, and Hubert G. H. Wilhelm, eds. *Barns of the Midwest.* Athens: Ohio University Press, 1995.
O'Neill, Maire. "Inventory of Historic Agricultural Structures in Gallatin County, Montana: A Pilot Survey." Unpublished paper, 1999, School of Architecture, Montana State University, Bozeman.
Palladino, L. B. *Indian and White in the Northwest: A History of Catholicity in Montana, 1831 to 1891.* Lancaster, Penn.: Wickersham Publishing, 1922.
Porter, Harry. *Barns of Illinois: A Pictorial History.* Virginia Beach, Va.: Donning Press, 2003.
Radford, William A. *Radford's Combined House and Barn Planbook.* New York: Radford Architecture Co., 1908.
———. *Radford's Practical Barn Plans.* New York: Radford Architecture Co., 1909.
Radojkovic, Jon. *Barn Building: The Golden Age of Barn Construction.* Erin, Ont.: Boston Mills Press, 2007.
Renewable Technologies. "Vaughn-Sun River Highway Project Montana Highway 89." Unpublished report, 1993, on file SHPO, Helena.
St. Ignatius Centennial Committee. *100 Years in the Flathead Valley.* St. Ignatius, Mont.: St. Ignatius Centennial Committee, 1954.
Sandoval, Judith H., T. A. Larson, and Robert Roripaugh. *Historic Ranches of Wyoming.* Casper, Wyo.: Mountain States Lithographing and Nicolaysen Art Museum, 1986.
Sanford, Dena. "National Register Multiple Properties Nomination for Korpivaara, Cascade County, Montana 1890–1945." 1991. On file MPA, Helena.
Sievert, Ellen, and Ken Sievert. *A Guide to Historic Lewistown.* Montana Main Streets Series. Helena: MHS Press, 1999.
Sloane, Eric. *An Age of Barns.* New York: Ballantine Books, 1967.
———. *American Barns and Covered Bridges.* New York: Funk and Wagnalls, 1954; reprint, Mineola, N.Y.: Dover Publications, 1982.
Smith, Phyllis. *Bozeman and the Gallatin Valley: A History.* Helena, Mont.: Two Dot Press, 1996.
Soike, Lowell J. "Within the Reach of All: Midwest Barns Perfected." In *Barns of the Midwest,* ed. Allen Noble and Hubert Wilhelm, 147–69. Athens: Ohio University Press, 1995.
Sommer, Robin Langley. *The Ultimate Book of Historic Barns.* San Diego, Calif.: Thunder Bay Press, 2000.
Stauffer, Charles, ed. *History of Beaverhead County.* Dillon, Mont.: Beaverhead County History Book Association, 1990.
Stephens, Henry, and Robert Scott Burn. *The Book of Farm Buildings: Their Arrangement and Construction.* London: William Blackwood & Sons, 1861.
Stevensville Historical Society. *Montana Genesis: A History of the Bitterroot Valley.* Missoula, Mont.: Mountain Press Publishing, 1971.
Stuart, Granville. *Forty Years on the Frontier*, vol. 1. Edited by Paul C. Phillips. Cleveland, Ohio: Arthur H. Clark Co., 1925.
Toole, K. Ross. *Montana, An Uncommon Land.* Norman: University of Oklahoma Press, 1959.
U.S. Department of the Census. *Census of Agriculture, 1900, 1910, 1920, 1930.* Washington, D.C.: Government Printing Office.
U.S. Department of Commerce, Bureau of the Census. *Ninth Census of the United States: 1870.* Washington, D.C.: Government Printing Office.
———. *Thirteenth Census of the United States: 1910.* Washington, D.C.: Government Printing Office.
———. *Fourteenth Census of the United States: 1920.* Washington, D.C.: Government Printing Office.

———. *Fifteenth Census of the United States: 1930*. Washington, D.C.: Government Printing Office.

Vaughn, Robert. *Then and Now; Or Thirty-Six Years in the Rockies*. Minneapolis: Tribune Printing Company, 1900.

Vaughn, Thomas, and Virginia G. Ferriday, eds. *Space, Style and Structure: Buildings in Northwest America*, 2 vols. Portland: Oregon Historical Society, 1974.

Vichorek, Daniel N. *The Hi-Line: Profiles of a Montana Land.* Helena: American & World Geographic Publishing, 1993.

Vlach, John Michael. *Barns*. New York: W. W. Norton, 2003.

Warhank, Josef J. "Fort Keogh: Cutting Edge of a Culture." Master's thesis, California State University, Long Beach, 1983.

West, Carroll Van. *Capitalism on the Frontier: Billings and the Yellowstone Valley in the Nineteenth Century*. Lincoln: University of Nebraska Press, 1993.

———. *A Traveler's Companion to Montana History*. Helena: MHS Press, 1986.

Worthen, Clifton B., and Oscar O. Mueller. "Central Montana." In *A History of Montana*, ed. Merrill G. Burlingame and K. Ross Toole, vol. 3, chap. XXII, 141–66.

Internet Sources

Fitzgerald, Deborah Kay. *Every Farm a Factory: The Industrial Ideal in American Agriculture.* New Haven, Conn.: Yale University Press, 2003. Accessed at yalepress.yale.edu/YupBooks/pdf/0300088132.pdf.

Frenchtown Historical Society. *Frenchtown Valley Footsteps.* Accessed at http://www.usbr.gov/dataweb/html/frenchto.html#N_7_.

James Manufacturing. *James Way*, Catalog 22. Accessed at http://www.antiquefarming.com/barn/dairy.html.

John Milner Associates. "Grant Kohrs Ranch National Historic Site, Cultural Landscape Report," 2004. Accessed at www.nps.gov/archive/grko/hrs/hrsf.htm.

Messiter, Charles Alston. *Sport and Adventure among the North American Indians.* London: R. H. Porter, 1890. Accessed at http//books.google.com, copyright bibliobooks.com.

Montana Department of Environmental Quality, Remediation Division. *Historical Context for the Bannack Mining District.* Accessed June 2005 at http://www.deq.state.mt.us/abandonedmines/linkdocs/ techdocs/4tech.asp.

Newby, Richard. "Reinventing Identity, Retaining Cultures, Discovering Lewis & Clark." Accessed February 10, 2010, at http://lewis-clark.org/content/content-article.asp?ArticleID=2975.

Texas Historical Association. "XIT Ranch." Accessed at http://www.tshaonline.org/handbook/online/articles/apx01.

Vogel, Robert. "Common Farm Barns of South Dakota, 1857–1958, Multiple Properties Document." South Dakota State Historic Preservation Office, Pierre. Accessed at history.sd.gov/preservation/OtherServices/SDBarns.pdf.

Archival Collections

Carbon County Historical Society. Barn Files. Red Lodge, Mont.

Central School Museum. Barn Files. Kalispell, Mont.

Frontier Gateway Museum. Archival Collections. Glendive, Mont.

The History Museum, Cascade County Historical Society. Barn Files. Great Falls, Mont.

Library of Congress, American Memory Project. Montana Collections. Washington, D.C. Accessed at www.memory.loc.gov.

Mansfield Archives. Archival Collections. University of Montana, Missoula.

———. Special Collections: Architectural Drawings. University of Montana, Missoula.

MonDak Heritage Center. Archival Collections. Sidney, Mont.

Montana Preservation Alliance. Barn Research Files, Helena. Files contain information from owners of all barns included in this book.

Montana State Historic Preservation Office. National Register Files. Helena.

Montana State University. Special Collections. Architectural Drawings. Bozeman.

Photographic Archives

Mansfield Archives. Photo Archive Collections. University of Montana, Missoula.

Montana Historical Society. Photo Archive Collections. Helena.

Museum of the Rockies. Photo Archive Collections. Montana State University, Bozeman.

Pioneer Museum. Photo Archive Collections. Bozeman, Mont.

Western Heritage Center. Photo Archive Collections. Billings, Mont.

Credits for barns, photographs, and illustrations

Unless otherwise noted, photographs are by Tom Ferris.

Institution abbreviations

GKR—National Park Service, Grant-Kohrs Ranch National Historic Site, Deer Lodge
MHS—Montana Historical Society Photograph Archives, Helena
MSU—Montana State University, Bozeman
MOR—Museum of the Rockies, MSU, Bozeman
PM—Pioneer Museum, Bozeman
SHPO—State Historic Preservation Office, Montana Historical Society, Helena
UM—Archives and Special Collections, Mansfield Library, University of Montana–Missoula

xiv: Barn courtesy Matador Cattle Company
4: Barn courtesy Jess Garfield
5: Barn courtesy Joanie Coffey-Taylor, Shaun Taylor, & Kristen Olsen
6: Notch-style details (clockwise from top left): Boyer Ranch, Galt Ranch, Bob Lydiard & Family, Rollins Barn–Ben & Mayme Ober, Galt Ranch, GKR, Switchback Ranch, Gilbert & Delores Myllymaki
8–9: Roof-styles left to right, (top) Franck Family; Blackwood Family; Gallatin Heights, LLC; (far left) Galt Ranch; (center) Norma Fitzgerald-Denny; John & Rose Brouwer; (bottom); Guercio Family, OW Ranch; Shimmin Family; (far right); Dailey Family
11: Courtesy Norma Fitzgerald-Denny
12: *The James Way: A Book Showing How to Build and Equip a Practical Up to Date Dairy Barn* (Fort Atkinson, Wis., 1917), 28
14: Barn courtesy Matador Cattle Company
16: Father P. J. De Smet, S.J., *Oregon Missions and Travels over the Rocky Mountains, in 1845* (New York, 1847), opp. p. 289
17: Kauffman, photographer, MHS, 941-068
18–19: L. A. Huffman, photographer, MHS, 981-779
20: (top) MHS, 947-221; (bottom) UM, 09-244
23: (top) UM, 72-242; (bottom) UM, 72-247
24: L. A. Huffman, photographer, MHS, 981-176
25: Library of Congress, (left) LC-USF-33-003187-M5, (right) LC-USF33-003291-M5
26: Courtesy U.S. Fish & Wildlife Service
28: (top) F. Jay Haynes, photographer, Haynes Fnd. Coll., MHS, Helena, H-3885; (bottom) MHS, 955-285
30: (left) Cornelius O'Keefe, "Rides through Montana," *Harper's Weekly*, 35 (October 1867), 579; (right) SHPO
31: SHPO
32–33: Barn courtesy Scott & Rita Tingey
34: Barn courtesy C. D. Didier Family Foundation
35: Barn courtesy Larry A. Schulz
36–37: Barn courtesy Thomas Nonemacher
38–39: Barn courtesy Tom & Fran Towle
40–41: Barn courtesy Robert & Sharon Snell
42–43: Barn courtesy U.S. Fish & Wildlife Service
44: Barn courtesy Bill & Marjorie Gardner
45: Barn courtesy Paula & Duane Johnston
46–47: Barns courtesy Glacier National Park
48–49: Barn courtesy Bar VW Ranch
50–51: Barn and historic photograph courtesy Bill & Helen Meadows
52–53: Barn courtesy Boyer Ranch
54–55: Barn courtesy The Home Ranch, Thomas R. Scheffer
56–57: Barn courtesy Chief Joseph Ranch, Shupp Family
58–59: Barn courtesy Patricia & Fred Wiele
60: Barn courtesy Fenton House
62: MHS, 940-695
64: MHS, Farms & Farming Coll.
65: (both) GKR, HS57, HS100
66: MHS, Farms & Farming Coll.
67: Albert Schlecten, photographer, MHS, Farms & Farming Coll.
68: PM, 94.81 p5423
69: MOR, 85.45.6
70–71: Barn courtesy Matador Cattle Company
71: (bottom) [M. A. Leeson], *History of Montana, 1739–1885* (Chicago, 1885), 283
72–73: Barn courtesy Jennifer Johnson, Gregory & Tracy Johnson
73: (bottom right) [M. A. Leeson], *History of Montana, 1739–1885* (Chicago, 1885), 989
74–75: Barn courtesy Matador Cattle Company
76–77: Barn courtesy Mogus Land & Livestock, Baty Family, Robert Parmenter Family
78–79: Barn courtesy Matador Cattle Company
80–81: Barn courtesy Matador Cattle Company
82–83: Barn and granaries courtesy Forrester Family, Ron & Rose Towery
84–85: Barn courtesy Marchesseault Family
86–87: Barn courtesy Rodoni Family
88–89: Barn courtesy William Tash Family
89: (bottom) [M. A. Leeson], *History of Montana, 1739–1885* (Chicago, 1885), 1017
90: Barn courtesy William Tash Family
91: Barn courtesy Montana Fish, Wildlife & Parks
92–93: Barn courtesy Linda Cue
94–95: Barn courtesy Madison County
96–97: Barn courtesy Harry Grose
98–99 Barn courtesy Hamilton Ranch
100–101: Barn courtesy Bradley Livestock
100: (top left) *Pioneer Trails and Trials: Madison County, 1863–1960* (Great Falls, Mont., 1976), 880
101: (top right) [M. A. Leeson], *History of Montana, 1739–1885* (Chicago, 1885), 785
102–103: Barn courtesy Whiterock Ranch
104–105: Barn courtesy Fenton House
106–109: Barn courtesy Jumping Horse Ranch
110–111: Barn courtesy John & Rose Brouwer
111: (top) Churchill-Amsterdam Historical Society, *A Goodly Heritage* (Bozeman, Mont., n.d.), 77
112: Barn courtesy Gallatin Heights, LLC

113: Barn courtesy John & Joyce Kamp
114–115: (top left and near right) Barn courtesy Blasberg & France Family
114–115: (bottom left and far right) Barn courtesy Bill & Edie Wright
116–117: Barn courtesy Mr. & Mrs. S. A. Taylor Family
118: Barn courtesy Venhuizen Family
119: Barn courtesy Joanie Coffey-Taylor, Shaun Taylor, & Kristen Olsen
120: (top left) Christine W. Brown
120–121: Barn courtesy Sieben Ranch Co.
122–123: Barn courtesy Bob Lydiard & Family
124–125: Barn courtesy Peter & Judy Burggraff
126: Barn courtesy Kleinsasser Family, Milford Colony
127: Barn courtesy Jeff & Marie Hoeffner
128–131: Barn courtesy Denis & Stacy Young
132–133: Barn courtesy Marks Ranch
134: Barn courtesy Ed Watson Family
135: Barn courtesy Joe & Terri Sologub
136–137: Barn courtesy Parini Brothers (John & Rudolph)
138: Barn courtesy Carlson Ranch
139: Barn courtesy Geary Bros. Ranch
140: (bottom) Detail, [M. A. Leeson], *History of Montana, 1739–1885* (Chicago, 1885), 555
140–141: Barns courtesy GKR
142–143: Barn courtesy Geary Bros. Ranch
144–145: Barn courtesy Robert & Britt Meyer
146: Barn courtesy Anderson Ranch Company; Joe & Leslie Tripp
148: MHS, 947-039
149: Barn courtesy The Bair Ranch Foundation
150: Fritz, photographer, MHS, 946-938
153: (bottom) Jan Bartel, photographer, courtesy Rebecca Hanna
154–155: Barn and historic photograph courtesy Hans Erickson Family
156–157: Barn and historic photograph courtesy Norma Fitzgerald-Denny
158–159: Barn courtesy Mack Bruno Sr., Marie Bruno
160–161: Barn courtesy Rossmiller Family
162: Barn courtesy Kalanick Ranch
163: Barn courtesy Jon & Sharla Tester
164–167: Barn courtesy J. C. Adams Stone Barn Company
168–169: Barn courtesy Kenneth Nevala family
170–171: Barn courtesy Gilbert & Delores Myllymaki
172–173: Barn courtesy Joe & Leslie Tripp Family
174–175: Barn courtesy Harry Mitchell Family
176–177: Barn courtesy Bear Cat Canyon Ranch, Hawn Family
178–179: Barn courtesy Campbell Family
180–181: Barn courtesy Jean Trammell Family
182: Barn courtesy MSU College of Agriculture, Montana Agricultural Experiment Station
183: Barn courtesy David Cripps
184–185: Barns courtesy Durl Gibbs/Cory & Clarajean Merrill Family
186–187: Barn courtesy Smith Family, Beaver Creek
188–189: Barn courtesy Galt Ranch
192–193: Barn courtesy The Bair Ranch Foundation
194–195: Barn courtesy Jess Garfield
196–197: Barn courtesy Eliasson Ranch Company
198–199: Barn courtesy Lloyd Shelhamer Family
200–201: Barn courtesy Doug & Ruby Bussell
202–203: Barn courtesy MSU College of Agriculture, Montana Agricultural Experiment Station
204–205: Barn courtesy Burns Family
206: (right) MHS, PAc 80-42
206–207: Barn courtesy MSU College of Agriculture, Montana Agricultural Experiment Station
208–209: Barn courtesy Robert & Penny Norman Family
210: Barn courtesy Page Whitham Land & Cattle, Michele Page
211: Barn courtesy Conrad Gustafson Family
212–213: Barn and historic photograph courtesy Kerstein Family
214–215: Barn courtesy Frank Hughes Family
216: Barn courtesy Huntley Project Museum of Irrigated Agriculture
219: L. A. Huffman, photographer, MHS, 981-471
220: L. A. Huffman, photographer, MHS, 981-1087
223–224: Courtesy Bones Brothers Ranch
226: (top) Courtesy Bette Kirscher
226–227: Barn courtesy Dailey Family
228–229: Anonymous
230–231: Barn courtesy Shimmin Family
232: Barn courtesy Joe & Millie Kimbro
233: Barn courtesy Rigler Family
234–235: Barn courtesy Susan Dusenberry
236–237: Barn courtesy Franck Family
238–239: Barn courtesy B Bar Ranch
240: Barn courtesy Bob & Jacky Shiplet Families
241: Barn courtesy Switchback Ranch
242–243: Barn courtesy Anna Belle Hill
244–245: Barn courtesy Hanson Family
246–247: Barn courtesy Switchback Ranch
248–249: Barn courtesy Jim & Sue Klessens
250–251: Barn courtesy Switchback Ranch
252: Barn courtesy Wegner Family
253: Barn courtesy MSU College of Agriculture, Montana Agricultural Experiment Station
254–255: Barn and historic photograph courtesy Harley O'Donnell
256–257: Barn courtesy Weldon Family
258–259: Barn courtesy Brett Ekle
259: (bottom) Courtesy John O. Miller Family & Sandra Miller
260: (bottom left) L. A. Huffman, photographer, MHS, 981-238
260–261: Barn courtesy Guercio Family, OW Ranch, Hanging Woman Creek
262: Barn courtesy Golden Willow Ranch, Rotten Grass Unit, Big Horn County
263: Barn courtesy Tim & Georgia Rugg Lanham
264–265: Barn courtesy Brown Cattle Company
266–267: Barn and historic photographs courtesy Bones Brothers Ranch
268: Barn courtesy Bud Forman, Forman Mizpah Ranch
269: Barn courtesy Fred (deceased) & Shirley Nunberg—Nunberg Farm
270–271: Barn courtesy Montana Department of Natural Resources & Conservation
272: (left) Evelyn Cameron, photographer, MHS, PAc 90-87.7-5
272–273: Barn courtesy Delbert Hoffer Family
274–275: Barns courtesy A. J. Pehl Family, George Ross Family, William Pehl Family
276–277: Barn courtesy Raymond Basta
278: Barn courtesy Deborah Oljar Klempel
279: Barn courtesy Daniel Young Family
280–281: Barn and historic photograph courtesy Patrick & Becky Wheeler
284: UM, 83.0104
285: (top left) Chere Jiusto; (top right) Teller Wildlife Refuge, Inc; (bottom) Chere Jiusto

Index

Italicized numbers indicate pages on which photographs appear.

H

I

J

N

O

About the Authors

Chere Jiusto has worked in the fields of Montana history and culture for over twenty years, most recently as the director of Montana Preservation Alliance. She is the author of *Montana Mainstreets: A Guide to Historic Hamilton* (Montana Historical Society Press). **Christine W. Brown** is an architectural historian with over ten years of experience documenting and evaluating rural properties. As outreach and education director at Montana Preservation Alliance, she promotes the preservation of Montana's historic buildings and places.

Tom Ferris maintains his own studio in Helena, Montana, but since 1995 has also worked as an archival photographer for the Montana Historical Society. Tom's work is in many private collections in the United States and overseas and has been collected by the Ministry of Culture in France and the Shimada Museum in Japan.